T0094110

Quantifying Counterfactual Military History

Forces shaping human history are complex, but the course of history is undeniably changed on many occasions by conscious acts. These may be premeditated or responsive, calmly calculated or performed under great pressure. They may also be successful or catastrophic, but how are historians to make such judgements and appeal to evidence in support of their conclusions? Further, and crucially, how exactly are we to distinguish probable unrealized alternatives from improbable ones? This book describes some of the modern statistical techniques that can begin to answer this question, as well as some of the difficulties in doing so. Using simple, well-quantified cases drawn from military history, we claim that statistics can now help us to navigate the near-truths, the envelope around the events with which any meaningful historical analysis must deal, and to quantify the basis of such analysis. *Quantifying Counterfactual Military History* is intended for a general audience who are interested in learning more about statistical methods both in military history and for wider applications.

Key Features:

- This book demonstrates how modern statistical techniques can measure the impact of counterfactual decisions.
- It examines the importance of counterfactual reasoning for both modern scholars and historical actors.
- It combines historical narrative, mathematical precision and data to create a straightforward presentation of both factual and counterfactual military history.
- It provides an original contribution to the debate over the validity and rigour of works of counterfactual history.
- It is written in a manner accessible to readers who have no formal training in History or Statistics.

Dr Brennen Fagan is a postdoctoral research associate supported by the Leverhulme Centre for Anthropocene Biodiversity and the Department of Mathematics at the University of York. He received his PhD in Mathematics, working with the York Historical Warfare Analysis Group to better understand human conflict by examining and modelling historical war data, and now studies the mathematics of biodiversity change.

Dr Ian Horwood is a historian at York St John University, where he is Senior Lecturer. His principal interests are in US military history, airpower history and the wars in Indochina. Dr Horwood received his PhD in History from the University of Leeds.

Professor Niall MacKay is a mathematician and theoretical physicist at the University of York. He has interests in military history, operations research and combat modelling. He received his PhD in Theoretical and Mathematical Physics from Durham University.

Dr Christopher Price is Senior Lecturer in History at York St John University. His main areas of interest are political, economic and military history in the twentieth century, especially British and US history in the period surrounding the Great Depression, the two World Wars and the Cold War. He received his PhD in History from the University of York.

Professor Andrew James (Jamie) Wood is a mathematician and systems biologist at the University of York. He specializes in the simulation and analysis of complex systems, and has interests across a range of modern international history and the analysis of warfare. He received his PhD in Theoretical and Mathematical Physics from Imperial College London.

ASA-CRC Series on
Statistical Reasoning in Science and Society

Series Editors
Nicholas Fisher, *University of Sydney, Australia*
Nicholas Horton, *Amherst College, MA, USA*
Regina Nuzzo, *Gallaudet University, Washington, DC, USA*
David J Spiegelhalter, *University of Cambridge, UK*

Published Titles

Data Visualization: Charts, Maps and Interactive Graphics
Robert Grant

Improving Your NCAA® Bracket with Statistics
Tom Adams

Statistics and Health Care Fraud: How to Save Billions
Tahir Ekin

Measuring Crime: Behind the Statistics
Sharon Lohr

Measuring Society
Chaitra H. Nagaraja

Monitoring the Health of Populations by Tracking Disease Outbreaks
Steven E. Fricker and Ronald D. Fricker, Jr.

Debunking Seven Terrorism Myths Using Statistics
Andre Python

Achieving Product Reliability: A Key to Business Success
Necip Doganaksoy, William Q. Meeker, and Gerald J. Hahn

Protecting Your Privacy in a Data-Driven World
Claire McKay Bowen

Backseat Driver: The Role of Data in Great Car Safety Debates
Norma F. Hubele

Statistics Behind the Headlines
A. John Bailer and Rosemary Pennington

Never Waste a Good Crisis: Lessons Learned from Data Fraud and Questionable Research Practices
Klaas Sijtsma

For more information about this series, please visit: https://www.crcpress.com/go/asacrc

Quantifying Counterfactual Military History

Brennen Fagan, Ian Horwood,
Niall MacKay, Christopher Price
and A. Jamie Wood

CRC Press
Taylor & Francis Group
Boca Raton London New York

CRC Press is an imprint of the
Taylor & Francis Group, an **informa** business
A CHAPMAN & HALL BOOK

Designed cover image: © Shutterstock, ID 251930104, Photo Contributor: Everett Collection

First edition published 2024
by CRC Press
6000 Broken Sound Parkway NW, Suite 300, Boca Raton, FL 33487-2742

and by CRC Press
4 Park Square, Milton Park, Abingdon, Oxon, OX14 4RN

CRC Press is an imprint of Taylor & Francis Group, LLC

ISBN: 978-1-138-59452-4 (hbk)
ISBN: 978-1-138-59238-4 (pbk)
ISBN: 978-0-429-48840-5 (ebk)

DOI: 10.1201/9780429488405

Typeset in Minion
by KnowledgeWorks Global Ltd.

Contents

About the Authors

Dr Brennen Fagan is a postdoctoral research associate supported by the Leverhulme Centre for Anthropocene Biodiversity and the Department of Mathematics at the University of York. He received his PhD in Mathematics, working with the York Historical Warfare Analysis Group to better understand human conflict by examining and modelling historical war data, and now studies the mathematics of biodiversity change.

Dr Ian Horwood is a historian at York St John University, where he is Senior Lecturer. His principal interests are in US military history, airpower history and the wars in Indochina. Dr Horwood received his PhD in History from the University of Leeds.

Professor Niall MacKay is a mathematician and theoretical physicist at the University of York. He has interests in military history, operations research and combat modelling. He received his PhD in Theoretical and Mathematical Physics from Durham University.

Dr Christopher Price is Senior Lecturer in History at York St John University. His main areas of interest are political, economic and military history in the twentieth century, especially British and US history in the period surrounding the Great Depression, the two World Wars and the Cold War. He received his PhD in History from the University of York.

Professor Andrew James (Jamie) Wood is a mathematician and systems biologist at the University of York. He specializes in the simulation and analysis of complex systems, and has interests across a range of modern international history and the analysis of warfare. He received his PhD in Theoretical and Mathematical Physics from Imperial College London.

CHAPTER 1

Could History Have Been Otherwise?

It is not given to human beings ... to foresee or to predict to any large extent the unfolding course of events. History with its flickering lamp stumbles along the trail of the past, trying to reconstruct its scenes, to revive its echoes, and kindle with pale gleams the passion of former days.

WINSTON CHURCHILL
eulogy for Neville Chamberlain, Hansard,
12 November 1940

HISTORY AND COUNTERFACTUAL HISTORY

Human life is contingent. The results of the smallest action are potentially momentous, and of the most dramatic act possibly insignificant. These consequences are extremely difficult to anticipate and can be uncomfortable or even frightening to contemplate. Life is thus not only contingent but appallingly contingent. The business of historians, however, is to make sense of the

DOI: 10.1201/9780429488405-1

past, to place the ideas and decisions of historical actors in proper context – to identify the significance of events and actions and to identify the momentum and direction of historical "forces", thereby counterbalancing contingency with some degree of necessity. There is little point in denying that this is a fiendishly difficult task, even with the past frozen in place like a prehistoric wasp in amber, and it is not, therefore, surprising that all historical perspectives are contested, with historians tending to separate into opposing camps over minor and major issues alike. Over time, dominant theories develop revisionist and later counter-revisionist schools. Warring historians would agree, however, that there is a truth to be revealed in their debate – even when they contend that they, rather than their rival, have perceived it.

To prove their case, historians appeal to evidence from the historical record – to what happened. Here, though, we encounter various difficulties. First, the constant realization of events provides us with a historical narrative, but the historian cannot possibly relate everything that happened. Good narrative history makes a choice of events to show their progression in a new light – to tell a good story. But the historian who wishes to do more than narrate must ask which events, which actions, are significant – which have long-term consequences – and let their answers determine their story. In doing so, they interpret the pattern of events, discern trends and interrogate the motives of persons and groups, aiming (even when this is only partly explicit) to arrive at a chain of causation. To some, the assassination of Archduke Franz Ferdinand may appear to have been an essential cause of the First World War, while to most it was merely the spark to a keg of gunpowder. But if a great war was inevitable, was this due to the inflexibility of railway timetables, the personalities of the statesmen or the converging and competing aims of the European powers?[1]

In writing history, it must always be remembered that a historical fact is simply one of numberless possibilities until the historical actor moves or an event occurs, at which point it becomes real. To understand the one-time possibility that became evidence, we must also understand the possibilities that remained unrealized. This is not a controversial statement, and historians attempt this synthesis as a matter of course. Indeed, to identify an event as important is to claim that in its absence or variation things might have turned out otherwise.[2] To the deep misfortune of the historical profession, however, attempts to refine our understanding and analysis of these possibilities have become mired in a spurious and damaging controversy concerning the validity or otherwise of "counterfactual history".

In the historical profession, the discussion of counterfactual history – a type of historical writing which creates an alternative historical narrative following an imagined difference in events at a given point in time – has generated more heat than light. At the extreme, it does little more than provide a battlefield for historians who dislike each other's positions, with the middle ground vacated in favour of lobbing shells across it from entrenched positions. Richard Evans, for example, accuses Niall Ferguson and a group of like-minded historians of pushing "wishful thinking" to create utopias or dystopias which chime with their political views and in some sense seek to validate them. Evans argues that this historical revisionism is symptomatic of the weaknesses inherent in historical speculation.[3]

Allan Megill goes some way towards clarifying this debate with a useful categorization that divides counterfactual history into two distinct types, the "exuberant" (which deals in chains of multiple alternative past events and outcomes which never happened) and the "restrained" (which considers only minimal and clearly possible changes to historical events). For Evans, for most historians and for us, "exuberant" counterfactual history is no more than implausible extrapolation, akin to playing a board game with history-shaped pieces. Yet Evans considers that even the more "restrained" counterfactual history also suffers from an inherent weakness, in that it doesn't go far beyond revealing what could be deduced within a more conventional historical framework anyway.[4] One solution to this conundrum is that the academic study of history needs to move beyond the customary analytic framework. Evans acknowledges, rather grudgingly, that historians necessarily indulge in counterfactual history whenever they are more than simply narrators of the historical record.[5] Explaining why history took the course it did inevitably involves a judgement on why other courses did not transpire. However, he argues that historical forces limit contingency to variations that can be encompassed within existing forms of historical analysis. As we will demonstrate, this is a questionable hypothesis, not least because it would stultify the development of historical methodology. As Tetlock and Belkin have shown, it is perfectly possible to create a cautious, careful, conservative qualitative methodology for dealing with contingency.[6]

Our purpose in this book is to take a few tentative steps towards using modern statistical theory and computer power to add a quantitative dimension to this debate – and to appreciate some of the dangers in doing so. We are interested in the role of chance in history, which means that we are equally interested in the probabilities of the events that happened – the facts – and their complement, the counterfactual. We will reflect on the implications for historians – not just for their writing of history but also for their views of the historical actors who made it. Before we set out our methodology, however, we need to discuss some connected topics – the existence of "watershed moments", the role of individuals in determining their outcome, and the special status of both of these in military history.

CRITICAL JUNCTURES AND "WATERSHED MOMENTS"

How should we select which events are appropriate for restrained counterfactual analysis? Later, we shall discuss the constraints imposed by the methodology available. But first, we consider how and why some events are more important than others. Such moments, more broadly including the actions of individuals and the development of institutions, are called "critical junctures" in modern political science.[7] We can never be certain of the existence of critical junctures, or of the grounds of their criticality, but "restrained" counterfactuals, if done with multiple perspectives and sufficient thoroughness, can surely make a distinctive contribution to the literature. Tetlock, Lebow and Parker's *Unmaking the West*, for example, considers whether the last few centuries' Western global dominance could be unmade by some minimal and perfectly plausible counterfactual changes to world history.[8] Such a "restrained" approach is very different from the piling of supposition upon supposition, with arbitrary choices made among multiple possible ramifications, that is characteristic of "exuberant" counterfactuals.[9]

If critical junctures exist, to what extent are their outcomes due to chance, rather than to perhaps unknown but in principle determinable sequences of decisions and events? The latter, and the extent to which they are inevitable outcomes of historical forces, are certainly historians' business, and Richard Evans is perhaps demarcating it when he observes that much recent counterfactual history succumbs to the temptation to see chance playing a role everywhere, "delight[ing] in emphasizing tiny causes for huge events".[10] To do so is clearly an abdication of the historian's responsibility at least to attempt to determine causes – even if, in the end, sensitive dependence on unknown hidden events may be indistinguishable from chance.[11]

To discuss this further, let's begin with a metaphor. Picture rain falling on a landscape. We can predict pretty accurately where most raindrops will end up – in the main river draining the basin in which they fall. But if the raindrop falls on the watershed ridge, its possible futures diverge and we have no predictability. Such simple critical junctures might equivalently be called "watershed moments". We are perfectly free to believe that these are very rare, that chance is greatly subordinate to historical forces. However, in the virtual landscape of possible historical realities – in the set of the many possible worlds on which the single droplet of our own world moves – watershed moments are not only conceivable

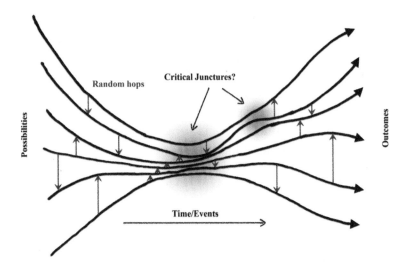

FIGURE 1.1 Critical junctures as regions in which historical trajectories are unusually dense.

but also almost inevitable, however rare they may be and however dominant the historical forces. Further down the hill, "these laws [of history] resemble rather those by which flood-water flows by hitherto unseen channels and forces itself finally to an unpredictable sea".[12]

The metaphor is easily broadened if we define critical junctures to be regions, illustrated in Figure 1.1, in which historical trajectories are unusually dense – "bottlenecks", towards which history converges and from which it diverges.[13] In passing through them, trajectories arrive from a broad set of pasts but separate into very different futures. They are moments of great contingency, the periods and places in which we find the "what ifs?" of history and the places in which hops from one trajectory to the next have the most effect. The largest, densest such regions would be associated with a sense of a culminating period, the end of an era, of an inevitable state of tension and uncertainty towards which all pasts tend.[14] In the rivers and seas of history, critical junctures are regions of strong currents and turbulence to which our boats are naturally carried, and from which their course is hard to determine and control. Yet still there is human agency: in Bismarck's words, "Man cannot create the current of events. He can only float with it and steer."[15] In the region of the critical juncture itself, there is unknowable entanglement and random hopping – that is, sensitive dependence on position and true randomness – and the two may be indistinguishable, either in principle or by historians. Such unknowability, and thereby randomness and agency in the hops, matters less before such a juncture, when trajectories are converging and necessity predominates in historical argument, and more afterwards, when trajectories are diverging and contingency dominates, while it matters most of all in the bottleneck itself, when trajectories are close-packed and about to diverge.

In this light, the opposition set up by Evans between contingency and inevitability, in which either all events are critical junctures or none are, seems rather false. Evans writes "the counterfactual proposes … 'If A had happened instead of B, then inevitably C, D and E would have followed … instead of X, Y and Z.' But, of course, a thousand other things might, or would, have intervened…".[16] If most details of history outside critical junctures really do not matter, then it is perfectly reasonable to consider contingency based on minimal changes (only) at critical junctures – or, if one prefers, to delineate the possible futures of a critical juncture without positing the changes or chances that could entail them. We would argue that the density-of-trajectories picture, which offers a continuous spectrum of criticality rather than a binary divide, is the natural worldview of the historian who does not believe that all history is reversionary and inevitable yet who does not wish to connect divergent outcomes with microscopic causes. There is much that can be said about turbulence around critical junctures without detailing the minutiae.[17]

But what are the laws shaping the landscape, its convergences and divergences? – can science help create a theory of critical junctures? Many writers on counterfactual history like to borrow ideas from science, but this can be rather dangerous, for science often uses words from common language which for the reader remain informal and intuitive but then co-opts them to specific, related and subtly different technical meanings. "Chaos" is one such word: its application as a concept from which to approach the crucial distinction between the stable and reversionary and the unstable and divergent is problematic and can be tendentious. Yet we do have to distinguish different types of instability and different forms and sources of apparent randomness.

In our raindrop metaphor, the historical forces are manifested in the shape of the landscape and thereby appear observable – even if only through watching the course of history's single droplet – and thus potentially tractable by conventional historical analysis. Most of the droplet's motion is reversionary, although it is unpredictable at the watersheds. The weather which produces the rain, by contrast, embodies a very different, (technically) chaotic kind of unpredictability: the weather is intrinsically unpredictable, and we cannot accurately predict rain even a few days in advance, still less where any one droplet will fall. But the general features and patterns of landscapes and weather systems do recur, just as only certain kinds of dynamics occur in "chaos theory", and only certain distinctive kinds of "catastrophe" – meaning sudden events caused by a change in the landscape – are possible.[18] Plenty of events which influence history appear in such patterns, structured yet unpredictable – earthquakes and stock-market crashes, for example.

A theory of critical junctures in history might therefore not be a quantitative mathematical analysis but rather a broader, more qualitative theory – something like our understanding of the weather before the advent of quantitative meteorology.[19] Critical junctures might occur of various types and degrees of density, and on varying time- and geopolitical scales. This is an enormous enterprise, but to say so is not counsel of despair. It is perfectly reasonable to hope to arrive on an empirical basis at natural timescales, scalings and cyclicity.[20] For example, if one believes that the onset of the First World War (or an equivalent cataclysm) was inevitable but that Franz Ferdinand's assassination was not the necessary or only possible trigger, a natural estimate of the time window might be a couple of decades, consistent with Randall Collins' timescale for geopolitical change.[21] Historical forces and predictability may be limited, but there is plenty we can say about the typical patterns we observe. History may not repeat itself, but it rhymes.[22]

THE INDIVIDUAL IN HISTORY

This can all seem a bit impersonal, and we need to restore some human agency. What of the historical actors themselves, moving in the eternal present? Contingency – uncertainty – matters deeply to them, for whom the future is unknown. It is surely a critical hindsight error to frame counterfactual analysis as a problem of significance only to later historians. It is clearly the case that historical actors have sought to interpret their circumstances in much the same way as do the historians of their future – that is, to calculate and implement the most advantageous course of action. In doing so, they faced much the harder task, fully aware of the possibilities of failure or misstep. From nineteenth-century military players of the *Kriegsspiel* to twenty-first-century policy makers using matrix games to attempt to work out the consequences of their decisions, their play-acting is conducted in deadly earnest with disbelief fully suspended, in the hope of reacting confidently to any possibility that becomes real and above all to avoid being surprised.

Standing somewhere behind such considerations, and far from fashion, is the idea of the "great man", the claim that the course of history is crucially dependent on the deep personalities of individuals.[23] Now, sometimes individuals clearly matter in history. Most historians, for example, are agreed that the character of the man Charles Stuart was one of the principal factors in the 1640s civil wars of the British Isles. So too was that of his nemesis Oliver Cromwell, loved by the English army for his tolerance of religious diversity (provided it was Christian and reformed). In Cromwell's case, one could even argue that through his beliefs his personality contributed to shaping the times – certainly Charles in contrast appears merely reactive. Yet Cromwell's beliefs were to a large extent shaped *by* his times, the "anthropomorphic cosmogony of the 17th century" in John Buchan's memorable phrase.[24] This is a crucial interplay, balancing Herbert Spencer's rebuttal of great man history on the grounds that "before [the great man] can remake his society, his society must make him" with Buchan's view that great individuals "are in a sense the children of their age, but they bring to their age more than they draw from it".[25] The former is the basis of Tolstoy's view of history, as described by Isaiah Berlin: not only are the conceits of "great men" empty, but nor also can we truly know, with any predictive power, the forces that shape us.[26] Yet personalities clearly differ, leaving of crucial importance the unanswerable question of whether the holding of power by the individual is an accident or rather is inevitable. It would be hard to argue that Otto von Bismarck's commitment to the avoidance of Europe-wide conflagration was an outgrowth of Prussian or even his new Germany's thought. Two generations later, whether the German Third Reich should be seen as the personal apotheosis of Adolf Hitler or as the culmination of German history can never be decided – but it is a striking counterfactual idea to combine the latter with a Nazi leader shrewder and more astute than Hitler, as Stephen Fry did in his novel *Making History*.[27]

MILITARY HISTORY

Military history provides plenty of examples of finely balanced outcomes whose consequences for wider history are much harder to determine. The Battle of Bosworth Field, fought in 1485 for the English crown, appears in retrospect to have been an epoch-making moment – the end of what we now call the Wars of the Roses, the beginning of the Tudor period, and perhaps the end of the Middle Ages in Britain and the beginning of the modern era. Of course, that's not how it seemed at the time when it was just another clash in the sequence of bloody battles that saw the extermination of large parts of England's aristocracy and commons alike. Contemporaries expected it to be succeeded by more such – it is only the survival on the throne of the battle's victor, Henry VII, that makes it appear otherwise to us now. If Richard III had won, would history look so very different? Probably the wars would have continued for longer and with intensified brutality. But underlying the ebb and flow of the wars were deeper currents – the Renaissance, the Reformation – which changed England and Europe and enabled the modern West to develop.

So, do the outcomes of battles matter? At the extreme, one might say "no, never" with the underlying inference that it is always the deeper currents of history which dominate in the end, that history reverts to its inevitable course.[28] Napoleon may have been turned back at the Battle of Aspern-Essling but the tide of war was inevitably heading towards his ultimate victory over the Austrians. At the other end is the "horse-shoe nail" argument, that some battle outcomes are both unstable (won on the toss of a coin, or the availability of a horse-shoe nail) and crucial in determining the geopolitical future – Waterloo or Gettysburg, perhaps.[29]

It's not difficult to think of ways to describe the spectrum of possibilities between the two extremes.[30] One would be to say that the distinction is just a matter of time. Then, in the former view, a battle's outcome may matter for a short while but not in the long run; in the latter, it may determine a society's long-term future. A middle ground is to guess that most battles' influence doesn't last very long. Towton, the bloodiest battle of the Wars of the Roses, may have been turned by snow and the wind direction, but its outcome mattered little a few decades later.[31] In the light of our earlier discussion of critical junctures, the nuanced view is that whether an event has long-term consequences is principally dependent on the context in which it occurs – a spark needs tinder if it is to start a fire.

Distinguishing the few battles which are forks in the road to different long-term futures is no easy task. The 1942 Battle of

Midway is a case in point. It was certainly highly contingent: for example, for want of a working catapult the Japanese failed to launch a sea-plane from the cruiser *Tone* that just might have discovered the American aircraft carriers in time – perhaps – to change the outcome of the battle. But would the Japanese seizure of Midway Island and the destruction of the American carriers have changed the outcome of the war in the Pacific and thus the long-term future of East Asia, or was the Japanese strategic misconception simply too great?

TOWARDS A METHODOLOGY FOR COUNTERFACTUAL MILITARY HISTORY

As Buchan noted, history is an art which is always trying to become more of a science.[32] It would be all too easy to give up hope of quantifying chance in history. But it would also be rather a shame, because there are a great many new techniques in statistics and probability ripe for use in historical analysis. We agree that there must always remain a role for incalculable chance and unknowable contingency in history. But science progresses and develops new modes of thought which, provided the effort is made, may have something to offer to historical analysis. Recent mathematics has opened up the study not only of dynamical change but also of chance, to such an extent that a great mathematician was able to call the turn of the millennium the "dawning of the age of stochasticity".[33] Nowadays, some chances can be calculated, and the likely distinguished from the unlikely.

Our purpose is not to propose a theory of history, or even a broad theory of critical junctures. Such attempts are worthwhile, perhaps timely, and can even point to forks in the road – but quantifying their probabilities is difficult. We concentrate on military history because it offers clearly defined moments of contingency in quantifiable contexts. We look at four scenarios of very different types, our understanding of which can be enhanced by the use of very different mathematical and quantitative techniques. We selected our cases because we considered them historically interesting and important, and because each was susceptible to a new quantitative approach either presently or by historical actors. They are all candidates for critical junctures, but we do not hypothesize about their long-term consequences, or whether these would have reverted to a convergent history in the long term.

But the capacity to quantify, in a novel and interesting manner, *is* critical. Quantified models have the particular virtue that, once they are in the open, everyone can discuss them: their simple logic is laid bare. Even if they are wrong, they provide a common basis for argument. If you think you do not have a model – well, you probably *do* have one, just not one sufficiently well specified that it can be subjected to precise argument.[34] Our models are not "black boxes", taking precise inputs and spitting out simple answers to historical questions in a manner insusceptible to interrogation. Rather they provide a new perspective, and a fresh jumping-off point for more conventional historical analysis. They can be particularly helpful in resolving the tension between factual and counterfactual, for, as noted earlier, to quantify the probability of an event that did not happen is also to quantify its complement, the probability of what did happen.[35] Whilst the fact that something happened might reasonably alter our *post hoc* estimate of its likelihood, there is a clear danger of going too far, of thinking that everything that happened was inevitable if only we could understand its causes well enough. Unlikely things *do* happen.

In this book, we explore various ways in which we can rethink some military engagements and decisions, quantifying the chances of what happened, what did not and what might have happened had different decisions been made. To the extent to which this is counterfactual, it is restrained rather than exuberant – as already noted, we won't pursue too far the implications for alternative histories. This approach is very close to wargaming; indeed, some of our computer simulations are like conducting simple wargames not a few times but many thousands or millions of times and examining the statistics of the outcomes. But we don't want to play wargames without understanding precisely the implications of their (necessarily) artificial rules, and the extent of the randomness and arbitrariness built into them.[36] Further, and without our intending to do so, we often found ourselves arriving at the importance of the individual in history. In several cases, our restrained counterfactuals added up to the mindset of a different individual making crucial decisions.

OUR CASES, CONSIDERED IN TURN

One way to approach the problem is to begin with a model for the way the battle is fought. This is a sub-discipline of operations research known as "combat modelling", but it begins with a very simple insight which emerged almost simultaneously in the United States, France, Britain and Russia before the First World War.[37] Now known as Lanchester's model, after its British inventor, it amounts to no more than the assumption that each of two forces engaged in battle causes damage in proportion to its numbers. Most analysts don't believe that this is true for much of warfare – it gives very different results from other models in which a battle is just the sum of individual duels, or long-range fire is unaimed – but if it were ever true, then it would have been so for the all-big-gun "Dreadnought" capital ships of the First World War at sea. It is relatively easy to play out the battles between such ships using the Lanchester model, but the crucial subtlety is in how to calibrate the parameters of the model, some of which we think we know and others we don't, to the historical outcomes. Suppose we do so, naively, and the outcome of the simulation is wrong – it does not match the historical outcome. Then how do we interpret our findings?

Nowadays, there is a methodology to tackle this, "Approximate Bayesian Computation" (ABC), which allows us effectively to distinguish two possibilities – that we have got the values of the parameters wrong, or that the historical outcome was unlikely.[38] This methodology has been around for less than 20 years, but it is already standard in science, where it can be used very effectively to challenge one's preconceptions with experimental data and then update them. As far as we know, ours is the first use of ABC for historical analysis.

In Chapter 2, we look at the largest clashes between dreadnoughts, the battles of Dogger Bank and of Jutland. Between them, they are a prime candidate for a critical juncture: as Churchill famously put it, the British commander at Jutland, Admiral Jellicoe, was the only person on either side who could (by losing British command of the sea) lose the war in an afternoon. Dogger Bank is a simple battle in which two lines of ships steamed on constant courses, firing at each other, whereas Jutland develops into a more elaborate dance of the lines of ships – simple calculus is rapidly replaced by subtle geometry. In reality, the Germans lost a ship at Dogger Bank and the British lost none, but we are able to show using ABC that this was more or less the reverse of the expected outcome – that the British got lucky. At Jutland, the opposite happened, with the British losing three battlecruisers, in part because the Germans learned the correct lessons from Dogger Bank, whereas the British were not forced to do so. Indeed, this tendency for the losers to learn the lessons of war better than the victors provides a natural balancing mechanism in military affairs. Jutland finished as a German numerical victory (in ships sunk) but a British strategic victory (in that Germany failed to realize Churchill's fear). Could the Germans, realistically, have turned their apparent tactical success into a strategic one? Did they come close? The answer isn't to be found in probabilities of sinking individual ships, but rather in the bigger picture – and this was very much determined by the mindsets of individuals. Jellicoe knew that he was fighting a battle he must not lose, much though the British public wanted a glorious win, and knew how to make the numbers sufficiently certain, while his subordinates Beatty and Evan-Thomas offer contrasting pictures of how an individual's dynamic can affect the course and outcome of a battle.

In Chapter 3, we move on to air war, and the Battle of Britain, Germany's 1940 attritional battle to gain air supremacy over south-east England. Again this is famously a candidate for a critical juncture, although it only truly becomes so when combined with a German invasion which, because of the failure to gain air supremacy, did not happen. How suitable is the Lanchester framework to assess this air battle? Although Lanchester originally conceived his model in the context of air war, it turns out that air combat is rather different – more like a set of duels, modified by some asymmetry between attacker and defender.[39] No simple model captures this, so we move instead to using statistical techniques which require no model – that is, which make no assumptions about the nature of air combat. We use this "bootstrap" technique, in which nothing but the actual daily air combat data are used, to provide a sample from which we resample with replacement. In essence, we are asserting that the daily data are all we know, and all we can ever know, about the battle: the only alternative histories of the battle that require no further assumptions are those in which the daily outcomes are precisely those of actual days in the battle. By extending or shortening phases of the battle, or changing the proportions of different targets, we can explore the effects of German decisions. Absolute numbers and probabilities are hard to reach and are entangled with other variables such as the weather, but the relative outcomes of different scenarios provide a basis for a better-informed discussion of the ensuing possibilities. The outcome is to quantify how counterfactual scenarios affected the big decision the Germans finally had to make, of whether to attempt the invasion of Britain.

The early chapters of this book employ modern analytical techniques to analyse the success or otherwise of decision-makers in the First and Second World Wars. In the second section of this book, comprising two chapters concerning the Vietnam War and nuclear deterrence in the Cold War, we reach a point where the advance of technology and technique convinced policy makers that they could harness mathematical method themselves, to disperse the fog of war. The superpowers of the time, but predominantly the United States, used analytical tools and computing power unavailable to their predecessors to assess evidence and guide current strategy and future actions – in each case making worse decisions as a result. This process introduces another crucial dimension so far overlooked in debate about counterfactual history. This is the extent to which historical actors place *themselves* in counterfactual situations by misinterpreting evidence and formulating policy on the basis of mistaken conclusions. The pressures of war and the imperatives of politics gave policy makers little time and space in which to think and impelled them to interpret data and the results of analysis in ways which reinforced their existing assumptions and their faith in the likelihood of desired outcomes. This was not necessarily because the information they received was bad. Mathematical analysis as a tool of policy-making was a process in its infancy and the sheer volume of data produced could be unwieldy and difficult to manipulate, but useful findings were often wilfully misinterpreted or ignored. As in most techniques of war, innovation was only likely to be successful after a chastening learning process. In both Vietnam and the nuclear arms race, the fog of war was slow to lift.

The Vietnam War, considered in Chapter 4, was the first to raise the possibility of fighting a war in which decisions would be based on computerized data analysis. Under McNamara's hyper-rational and logistical mindset, and in the first flush of computer power, the United States collected vast quantities of data about the state of every hamlet or village and every military unit in Vietnam. For us, the benefit of hindsight is facilitated by vastly greater computer power – and with modern data analysis techniques, we can see that such decision-making could never have been well-founded, for the data set itself provides at best weak support for any course of action. But the crucial question which transcends hindsight is whether contemporary data analysis could have enabled a better decision to be made, and how this could have been achieved. The best results follow when qualitative analysis, mathematical models and data analysed with caution and understanding come together[40] – and here, there are indications that a strategy of pacification (as opposed to a war of attrition) could have been comprehended, implemented and yielded better results, perhaps even a victory – but, as so often, the contingency turns out to be far more on personalities and politics than on either the quality of the analysis or the military balance.

Chapter 5, which deals with the applied theory of nuclear deterrence culminating in the "Able Archer" nuclear crisis of 1983, resonates most closely with Chapter 4 in that the same actors in the post-1945 environment were attempting to base policy on mathematical analysis, in this case predominantly game theory. Unlike the Lanchester models of battleship engagements presented in Chapter 2, game-theoretic models of nuclear war applied by policy advisors in the United States were not up to the task of conceptualizing the realities of the nuclear balance. For most of the early Cold War, this was not necessarily a dangerous state of affairs due to an asymmetry of information. Soviet leaders were fully aware of their state of inferiority relative to the United States, which was so acute that they were unable to produce any real nuclear threat to the continental United States until the 1960s. The Americans, however, could only guess at the state of affairs in the USSR and wildly overestimated Soviet strength, aided by astute Soviet propaganda and the self-interest of the US military-industrial complex in reaching the same conclusion. This state of affairs produced security based on the illusion of strategic balance, which in Soviet eyes saved them from the pre-emptive attack favoured by some in the United States had their true weakness become known. It is difficult to say whether this was a practical success for game theory or not, but a situation in which each of the protagonists mistakenly ascribes aggressive capability and intent to the other is obviously perilous. Game theory, even in its modern probabilistic variants, depends crucially on assumptions of rational actions and objectives and is at its most dangerous precisely when we do not appreciate our opponent's state of mind, and our consciously rational responses are based on misconceptions and will thus themselves appear irrational.

Such a situation occurred in the period of the so-called Second Cold War, which broadly encompassed the breakdown of *détente* in the mid-1970s through to the assumption of power by Mikhail Gorbachev in the USSR in 1985. In this period, both sides thought the other to be approaching an ability to win a nuclear war with an overwhelming first strike and each came to believe that the other might intend to deliver it. We introduce a new "game" in order to tell the story of these misperceptions and their culmination in the Able Archer Crisis of 1983, in which an ill-advised American operational plan to deter the Soviets by threatening to "decapitate" their leadership with a highly accurate first strike interacted with the Soviets' rational fears and intelligence failings to produce an unstable deterrence dynamic which mirrored previous Cold War episodes. The idea of our game is to simplify modern notions[41] by tensioning misperceptions of an opponent's probability of a nuclear first strike against one's own rational response to it. We will see that this can easily lead to disaster. The overall lesson, perhaps, is that opponents should always talk to each other; even with the attendant disinformation and deceptions, it is always less dangerous, and more helpful to one's own interests, to try to achieve some mutual awareness and understanding.

The object of this book is not to reinvent the past and imagine alternative futures. Rather we have used a mix of quantitative techniques and mathematical concepts – some established in this field, some not, but each appropriate to the limited context in which we apply it – to present a fuller picture of the dynamism of the historical process, acknowledging the pressures and uncertainties faced by historical actors contemplating their future and how they might shape it. Yet, in doing so, a recurring and unexpected theme is that our simple, quantitative, restrained counterfactuals often combine to lead us to individual commanders, their personalities and the ultimate grounds for their decisions. For success in the early stages of the battle of Jutland, we find that the British battlecruisers need three restrained counterfactuals: more gunnery practice, better flash discipline and better command to ensure integration of the fleet. Three "restrained" might add up to one "exuberant" counterfactual – yet all were in the gift of the British commander David Beatty. For German success in the Battle of Britain, different operational and strategic conceptions were needed from Hermann Goering and Adolf Hitler. The same was true of Johnson, Nixon, McNamara and Westmoreland in Vietnam, and Reagan and Andropov in Able Archer: replace them with different personalities and different goals, and different outcomes become much more plausible. Without its having been our intention, we found ourselves agreeing with Buchan on the importance of "sheer force of personality and mind".[42]

Our introductions of scientific and mathematical techniques – often contemporary in origin with the events under discussion but enhanced by present-day analysis and computer power – are there to reinforce our understanding of probability and occurrence. To support the reader, we provide an appendix of some helpful mathematical background to accompany our cases; this can be consulted as the reader's curiosity requires. But our final aim is always to return to historical analysis, to explore the implications for historians and historical actors. We do not claim that such techniques necessarily make for better history than can be achieved without them, but we do believe that they can offer a different perspective, allowing the lights and shadows to fall in new ways. This study has taken us in directions which are not common in academic collaboration, but which we hope will demonstrate that collaborative research exploring what had been dead ground between the sciences and the humanities is long overdue.

NOTES

1. The railway-timetable argument is famously due to A. J. P. Taylor, *War by Timetable*.
2. A classic study of the effects of railroads on American economic growth considers how the economy would have developed in their absence. Fogel, *Railroads*; see also McCloskey, "Counterfactuals".
3. Richard Evans, *Altered Pasts*, and Niall Ferguson, *Virtual History*.

4. Allan Megill, *The New Counterfactualists*. Evans and Ferguson are critiqued in Tucker, *Historiographical Counterfactuals*, and in Sunstein, *Historical Explanations*. For commentary on counterfactual history, see Bunzl, *Counterfactual History*, and Chs.1 and 12 of Tetlock, Lebow and Parker, *Unmaking the West*. See also E H Carr's classic *What is History?*, Chap. 4, "Causation in History" – for Carr, counterfactual history was a mere "parlour game" (p.97). For the role of counterfactuals in politics, see Fearon, *Counterfactuals and hypothesis testing*.

5. Evans, *Altered Pasts*. See, for example, p35, pp50–51 and p67.

6. Tetlock and Belkin, *Counterfactual Thought Experiments*.

7. Capoccia and Kelemen, *The Study of Critical Junctures*. The science fiction writer Harry Harrison called such critical junctures "alpha nodes" in his novel *Tunnel Through the Deeps*, in which a counterfactual failure of the Moors to win control of Spain at the Battle of Las Navas de Tolosa in 1212 led to a very different next 700 years. Nevertheless in his novel the British Empire emerged and achieved an even greater degree of pre-eminence than in reality. John Buchan, in his 1929 lecture *The Causal and the Casual*, also makes the crucial distinction between the many small variations which do not have long-term consequences and the few which do (p21).

8. Tetlock, Lebow and Parker, *Unmaking the West*. The tentative answer which emerges after considering various restrained counterfactual possibilities is "before the Reformation, probably; afterwards, probably not."

9. The distinction between "exuberant" and "restrained" is much like the distinction between science-fantasy and an older form of science fiction in which some small change in the science allows a very different world to develop – as, for example, in John Wyndham's novel *Web*, in which spiders are physically unchanged but mutate to become social animals.

10. Evans, *Altered Histories*, p35. That chance plays an essential role in history is also a natural position for historical novelists to adopt – it is hardly surprising that Buchan, one of the best of them, argued against historical inevitability in *Causal and Casual*.

11. The tension between true "aleatory" – dice-playing – randomness and the apparent equivalence of unknown or even unknowable hidden causes is perhaps best worked out in quantum mechanics, where it is now understood that the theory's apparent randomness cannot be replicated by local hidden determinate variables.

12. Taylor, *Bismarck*, p70. For an argument against "critical junctures", see Collins, *Turning Points and the Uses of Counterfactual History*. The tension between stable and unstable, reversionary and divergent, is not new: see Fisher, *Modern Historians*, whose argument might be paraphrased as "history is not geometry but neither is it romance". The point about basins of attraction within a watershed is made in Bonneuil, *Mathematics of Time in History*.

13. The picture of contingency as divergent trajectories, sensitive to initial conditions, and necessity as convergent trajectories, reverting to a common future, was introduced by Yemima Ben-Menahem, *Historical Contingency*. Our concept of critical junctures as regions of unusually dense trajectories is a combination and extension of her notions. The "bottleneck" metaphor is developed and criticized by Collins, *Turning points*, which also introduces the notion of "pseudo-turning points", following which history temporarily diverges but then reverts on his geopolitical timescale of 30–50 years.

14. Capoccia and Kelemen, in *The Study of Critical Junctures*, propose a formula to quantify the size of critical junctures, which combines (for specific outcomes) probability jump with temporal leverage. Our analogue of the probability jump would be to take the logarithm of the density of trajectories at the juncture divided by its typical value, which is then related to the notion of "surprise" (see appendix).

15. Taylor, *Bismarck*, p70. The metaphor is echoed by Braudel: "the true man of action is he who can measure most nearly the constraints upon him, who chooses to remain within them and even to take advantage of the weight of the inevitable, exerting his own pressure in the same direction. All efforts against the prevailing tide of history – which is not always obvious – are doomed to failure" (quoted in Collins, *Turning Points*).

16. Evans, *Altered Pasts*, p81.

17. Tuchman's *August 1914* and Watt's *How War Came*, dealing with the outbreaks of the First and Second World Wars, respectively, are exemplars.

18. Complex changes, in which the governing system itself alters and thus the patterns of stability and instability are varied, are called "bifurcations" and can cause "catastrophes", in which a hitherto stable system becomes unstable. Such "Catastrophe Theory" (and the closely related notion of "chaos") was perhaps oversold following its establishment in the 1960s by René Thom, but its mathematical certainties do relate to the real world: when change occurs, it does

so in one of a range of classifiable ways. See Poston and Stewart, *Catastrophe Theory* and Gleick, *Chaos*. Reisch, *Chaos, History* argues that history is chaotic; Roth and Ryckman, *Chaos, Clio*, counter that it is not.

19. The mathematization of meteorology was largely due to one man, the physicist and Quaker pacifist Lewis Fry Richardson – who, perhaps coincidentally, was one of the first scientists to theorize mathematically about war, in his books *Arms and Insecurity* and *Statistics of Deadly Quarrels*.

20. A broad such theory is provided by Peter Turchin, *War and Peace and War*. Here, shorter "fathers and sons" cycles of civil war are accompanied by longer term cyclicity in economic history – most famously "Kondratieff waves", dealt with in Turchin, *Secular Cycles* – and these are set against the rise and fall of civilizations over millennia. This in turn is due to the growth of a society's coherence ("asabiya") and exceptionalism (itself due to the multiplicity of threats it faces) and their subsequent decline.

21. For Collins, geopolitical change is naturally resolved only down to periods of about 30–50 years, and we should not expect greater precision than this in geopolitical prediction. In our picture, this would be the typical duration of the largest critical junctures. Collins, *Turning Points, Bottlenecks and Fallacies*.

22. George R. R. Martin's *Song of Ice and Fire* series (filmed as *Game of Thrones*) is then a defensible extreme of exuberant counterfactual history. *Game of Thrones* takes place in a fantasy world, not connected by any specifiable set of changes to our own, which appears stuck in the Middle Ages: it has experienced no Renaissance, no Enlightenment. In it, the forces and patterns – the "weather" – of mediaeval history play out in internecine strife loosely inspired by the Wars of the Roses, and its undoubted fiction gives licence for exploration of historical forces free from concerns about alignment with the historical record.

23. "Great man" theory runs most of its course in the nineteenth century, placed front and centre in Thomas Carlyle's *On Heroes*…. But the strength-of-personality point was embedded by Carlyle in arguments that could be seen as antecedents of twentieth-century fascism: see Schapiro, "Carlyle, prophet of fascism".

24. Buchan, *Oliver Cromwell*, p446. Buchan's point is that we can never understand the shaping of Cromwell by his times: "For a modern man that is impossible. The narrow anthropomorphic cosmogony of the seventeenth century is gone."

25. Spencer, *The Study of Sociology*. Buchan views such arguments as "undue simplification" by the "would-be scientific historian" (*Causal and Casual*, p14–16).

26. This position is famously explored by Isaiah Berlin in *The Hedgehog and the Fox*. One could argue that it is a great irony of history that both aspects of Tolstoy's view were exploded by the events of the Russian revolution, in which Lenin's individual actions were crucial in securing the success of the Bolsheviks and thereby realizing Marx's supposedly ineluctable historical forces. A scientific analogy is made by Turchin: Tolstoy's position, in which only a society of multiple interacting individuals can shape history, is that of "statistical physics", which deals for example with how laws for gases emerge from the kinetic behaviour of individual molecules (*War and Peace and War*, Ch.12). Another analogy, from biology, is our recent understanding of the behaviour of flocks, swarms and herds, and how they can follow from simple individual choices. See Wood and Ackland, *Evolution of the Selfish Herd*.

27. Fry, *Making History*. A classic but rather extreme example of this view is Taylor, *Course of German History*. We delight here in complying with Herzog's corollary to Godwin's Law: "any discussion of counterfactual history will eventually turn to Hitler or Nazi Germany as an example". Herzog, Ch.13 of von Dassanowsky, *Quentin Tarantino*.

28. It was such a view that Buchan railed against in *Causal and Casual* – or, to be more precise, that "pseudo-scientists... believe that they can provide a neat explanation of everything in the past by subsuming it under a dozen categories" or, like Marx, "into the iron bed of a single formula" (p43). See also Isaiah Berlin, *Historical Inevitability*. A satisfying resolution is the *longue durée* view of the *Annales* school, which sees warfare and politics as merely the tectonics of plates moving on the surface of the deeper movements of social, geographical and technological history. See, for example, Fernand Braudel, *History of Civilizations*.

29. For both these cases, see Cowley, *What if?* The proverb has multiple variations, some apocryphally associated with the Battle of Bosworth Field. A classic form is *For want of a nail the shoe was lost; For want of a shoe the horse was lost; For want of a horse the battle was lost; For the failure of battle the kingdom was lost – All for the want of a horse-shoe nail.*

30. And it is of course essential to accept that such a spectrum exists – that counterfactuals can matter somewhat and sometimes rather

than either always or never. The rejection of counterfactual history on the false ground that it is a binary choice is made by Tristram Hunt (*Pasting over the Past*, quoted in Evans, *Altered Pasts*, p50): "implicit in ['what if' history] is the contention that social structures and economic conditions do not matter".

31. Henry's survival after Bosworth can in large part be attributed to his own talents. In contrast, the limited influence of earlier battles such as Towton appears due at least somewhat to the previous Yorkist monarchs' inadequacies. For an excellent summary, see Bindoff, *Tudor England*.

32. Buchan, *Causal and Casual*, p7.

33. Mumford, *Dawning of the Age of Stochasticity*.

34. This argument is cogently and concisely made in Epstein, *Why Model?*

35. An interesting experiment in asking historians to quantify their beliefs as probabilities is recounted in Tetlock and Lebow, *Poking Counterfactual Holes*.

36. For literature and a review of the dangers of digital wargaming, see Rex Brynen, *Virtual Paradox*.

37. Lanchester, *Aircraft in Warfare*; Chase, *A Mathematical Investigation*; Baudry, *The Naval Battle*; Osipov, *Influence of Numerical Strength*.

38. ABC was introduced by Marjoram, *MCMC without Likelihoods*, building on a large body of previous work.

39. Horwood, MacKay and Price, "Concentration and Asymmetry in Air Combat".

40. As exemplified by the work of Seymour Deitchman, discussed later.

41. For example, of "subgame perfection" and of "games with incomplete information" (Bayesian games).

42. Buchan, *Causal and Casual*, p15.

CHAPTER 2

Could the Germans Have Won the Battle of Jutland?

FIGURE 2.1   A Second World War photograph of the American battleship *USS New Jersey "taking it green" late in the war. This photograph is taken as the US Fleet navigated through a Pacific typhoon that did considerable damage to the fleet. USS New Jersey was the second of the USS Iowa's class battleships that remained in service until the 1990s.* (USN Naval History and Heritage Command, image 80-G-291407)

DOI: 10.1201/9780429488405-2

HISTORICAL CONTEXT

For twenty years before the First World War, Britain and Germany had been engaged in an arms race. One of its most striking technological manifestations was HMS *Dreadnought*, a fast, heavily armoured, all-big-gun battleship, and by 1915 the two nations had built over 50 of them. In parallel to the shipbuilding programme was the development of ideas about how best to use them – and, overhanging all of this, the human factor, the qualities of the commanders. There were only two battles between dreadnoughts, a smaller action at Dogger Bank in 1915 and the larger clash at Jutland in 1916. Above all Britain could not afford to lose, and thereby lose its command of the seas. As Winston Churchill said, the British commander John Jellicoe was "the only man on either side who could lose the war in an afternoon".[1] Could this have happened?

One of the purposes of this book is to discuss how mathematical modelling can shed light on messages from history that are perhaps unclear before quantitative analysis. With this in mind, all mathematical modelling starts with a set of assumptions, so here is one to begin with: that the sea is flat. For the purposes of understanding interactions between battleships several hundred metres long and firing shells up to a few tens of kilometres, this is pretty accurate. We don't need to worry about the curvature of the earth, and whilst there are plenty of accounts of even battleships "taking it green" on the sea (Figure 2.1) – there have been no instances during modern battles of sea state making differences to the positioning of the ships.[2] Further, the sea surface is homogeneous and isotropic. Unlike in the age of sail, it is reasonable to say twentieth-century battleships could move in any direction with the same ease, and (at least in the open ocean) all positions were alike.

Why state this obvious fact? The first job in any quantitative modelling exercise is to understand which details have an important effect on the outcome – which are relevant and which are not.

For powered ships at sea the ability to describe the battle with high accuracy as occurring on a homogeneous 2D sheet, rather than in a 3D space (as with air combat) or on a lumpy and varied landscape (as with ground war), is a huge conceptual advantage. For mathematics, this is the first of the "rules of the game", the title of Andrew Gordon's book on the tactical uncertainties which beset the naval powers, particularly the British, for the hundred years from the Battle of Trafalgar to the outbreak of the First World War.[3]

COMMAND OF THE OCEAN

To set the scene, in 1812 it was unambiguously clear that the Royal Navy was in complete command of the ocean.[4] With wooden ships, sails and cannon, the Navy knew what to do, and it had the ships and commanders to do it. This clear dominance also reflected Britain's industrial and technological ascendancy. This would be challenged over time as ships changed: wood was replaced by steel, sail by steam turbine and cannon by large-calibre breech-loading rifles firing high-explosive shells. Of course, this was not a smooth progression: HMS *Warrior* now seems an ungainly hybrid of wood and iron, but its sail/steam combination was common at the time. The peculiar duel between USS *Monitor* and CSS *Virginia* at the Battle of Hampton roads in 1862, in which the two ironclad ships fought for nearly four hours with little effect on either, illustrates the consequences of imbalance between offence and defence for an effective navy.[5] For a brief period this imbalance was so great that the ancient technique of the ram was brought back into use; indeed it sank ships at the battle of Lissa, where in 1863 the Austrian Navy fought the Italians in a short battle between forces with a mixture of wood and iron, steam and sail. Lissa raised the importance of the ram in battleship design for the next 50 years and caused a generation of commanders to overemphasize the importance of closing the range in their desire to repeat the success of the Austrian Admiral Tegetthoff. Yet not a single major surface ship was sunk in battle by a ram ever again.

By the 1870s, the British muzzle-loading rifle could penetrate the compound armour of the time, and the balance between offence and defence see-sawed throughout the period. One of the resulting uncertainties concerned what the calibre of the main gun should be. A 12″ gun fires a shell eight times heavier than a 6″ gun, but the gun itself weighs eight times more, so a ship can mount eight times as many 6″ guns, balancing these effects. Further, the 6″ gun can fire at a much faster rate, so that a battery of 6″ guns gives a much higher delivery rate of weight of shell. However, most of this is wasted if a 12″ shell can penetrate the enemy's armour but a 6″ shell cannot. Ultimately it was the critical necessity of armour penetration which proved decisive for naval construction. Larger and larger guns appeared during the late nineteenth century.

An event highlighted in Gordon's book is the accidental sinking of the HMS *Victoria* in the Mediterranean in a collision with HMS *Camperdown* in 1893. This event, at least temporarily, resolved a crucial debate within the Royal Navy regarding central versus distributed command, in favour of the former.[6] Vice-Admiral George Tryon, the fleet commander who perished with his flagship, was ironically a strong advocate of reinvigorating the Nelsonian tradition of independence of ship's captains. *Victoria*'s peculiar configuration of gunnery, a single turret with a pair of 16 1/4″ guns, was not untypical of the design of warships of this time despite the imbalance this caused,[7] illustrating the flux of opinion on the best gun configuration. The evolution of this debate would be critical to the performance of the Royal Navy in the two World Wars to follow, but it is perhaps striking that these conversations initially took place without reference to the importance of the interplay between gunnery and command. These constraints would be in sharp focus once the potential of the big-gun ships that would emerge in the subsequent decades was realized.

The consequences of getting the gunnery "right" were first illustrated by the events in the Tsushima channel between Japan and South Korea in May 1905. A Russian fleet, sailing half-way around

the world from its base in the Baltic, attempted to relieve a belea-
guered squadron at Vladivostok during the Russo-Japanese War.
The Russian fleet was decisively beaten, losing nearly 30 ships,
including all eight of its modern battleships, thereby decimating
its naval power as well as Russia's status. It had a forebodingly
important effect on Japanese naval philosophy, a decisive naval
encounter, a *kantai kessen*, thereafter being seen as the only route
to victory. Tsushima was also the first demonstration of the superiority of accurate large-calibre gunnery over batteries of smaller
weapons.

The era of naval uncertainty was – unambiguously, in hindsight –
brought to an end by the launch of the all-big-gun HMS *Dreadnought* in 1906. The adoption of the all steel-armoured battleship,
powered by steam turbines and with as many big guns as it could
carry in turrets, marked the beginning of a breathless era of naval
development which lasted until the Washington Naval Treaty of
1922. Admiral John ("Jacky") Fisher[8] oversaw enormous technological change during this period, driven in part by his tactic of
"plunging", the relentless desire to render existing ships – even in
his own Navy – obsolete by new developments. Indeed, the game-
changing *Dreadnought* would arguably be rendered obsolete not
just once but perhaps as many as three times – in gunnery calibre, speed and armour – between its launch and the clash of the
British and German fleets at Jutland in 1916.

Whilst the outbreak of war in 1914 was to take many by surprise, the naval arms race was firmly embedded. Admirals in both
Britain and Germany had large fleets both in being and in creation. Around the world, in Russia, America, France and Britain,
commanders and captains pondered the same questions. How
should they direct their fleets? Should they be autonomous, or
should they use flag signals or even wireless sets, whose future
importance had been foreshadowed at Tsushima, to attempt
fleet-level control? What was the best balance among armour,
speed and firepower? Should ships be fast and thinly armoured
or should they be slow, well-protected gun platforms? And, given

limited resources (even for the Royal Navy), which is better, one ultimate ship or three inferior ones?

QUANTIFICATION

These conundrums, and the clearly demonstrated penalties for failure, sparked much discussion in the literature of the time. We will briefly discuss each in turn, but the discussion can usefully be framed in terms of the counterfactuals. The unrealized alternative answers to these questions are perfect illustrations of the restrained counterfactuals which are characteristic of wargaming in general and represent the situation faced by historical actors who cannot see their future but need to choose between actions which will shape it. Given a set of possibilities, under constraints, what is the best way to proceed? What are the different outcomes of adopting a particular choice, and are they better or worse than other realizable alternatives? Of course, we could proceed solely through verbal discussion, but this only tends to reveal different opinions, different schools of thought. In Germany, such discussions were framed within the Prussian *Kriegsspiel* wargaming tradition, which provided evolution towards a quantitative description, but not a complete one. Quantification of the problem provides at least one new dimension, and possibly some resolution.

Early in the *Dreadnought* era, several authors began constructing a more mathematical description to enable a rational *calculus*-based counterfactual to be constructed. Baudry, working in France and published in 1910, understood the essence of what became known as the "Square Law" in the 1913–1914 work of the British engineer Frederick Lanchester. The same ideas were derived by Osipov in Russia – although his work did not become known in the West until much later – and in classified 1902 work by J. V. Chase for the US Navy. Perhaps the easiest and simplest to describe is the work of Bradley Fiske. Writing in a Prize Essay of the US Naval Academy, he uses no differential calculus but rather provides a set of tables – now we would recognize them immediately as spreadsheets – which describe the attrition of the fighting strength of two forces engaged with each other. His conclusions note the disproportionate value of a short initial period ("five minutes") of unopposed fire, alongside the growth of any initial imbalance between the two forces.[9]

The central assumption is that the two sides each do damage to the opposition in proportion to their own numbers but independent of enemy numbers. Such simplicity rarely holds in war, but if it were ever so, then it was in big-gun naval engagements: hits will be proportional to the number of guns, whose target is unambiguous and against which there is no effective defence. The implication for the process of attrition is that the damage you do not only degrades your enemy but also degrades his ability to hit back. The reader can easily illustrate this by playing the traditional game of battleships in two different ways: either with one shot per turn, or with a number of shots equal to the number of remaining friendly ships (sometimes called the "salvo variant"). In the latter case, any imbalance is rapidly exaggerated, as Fiske noted. If instantaneous damage is proportional to numbers then when this is summed the result is a square of numbers giving Lanchester's "Square Law". Trajectories are hyperbolae, diverging as time progresses, and each force's capacity to win the battle is proportional to its units' effectiveness multiplied by the square of their numbers. Whilst ostensibly devised by Lanchester for aerial combat, he illustrated it with the battle of Trafalgar, and the basic concepts were well embedded in naval thinking by the outbreak of war. Indeed, the British Admiral Jellicoe was later to write to Lanchester saying that "your N-squared law has become quite famous in the Grand Fleet".[10] We will refer to the body of such thinking as *Lanchestrian*.

If the *calculus* of naval combat emphasized the importance of attrition and of firing without receiving fire, then the goal of fleet *geometry* was to enable the entirety of one's fleet to engage the enemy, thus enabling Lanchestrian concentration and its associated calculus to be achieved. British staff officer Reginald Drax was obsessed with the idea: "It is 'Applied Geometry' that must ensure for us the crushing effect to be obtained by bringing all our forces into action at the same moment. 'Geometry... Geometry... Geometry...'. The leader of a large fleet should diligently cultivate in himself a 'geometric sense'".[11] But what geometry was required, and what were the best control strategies to achieve it? Drax's papers demonstrate the level of thought put into these questions. For example, the benefits of turning in sequence – that is, each ship turning at the same place and continuing in the same line in a different direction – relative to turning simultaneously (and thereby, in a full "U" turn, reversing the order of the ships) formed a crucial part of this discussion. The former maintains a clear level of central command but increases the danger to ships under fire, whilst the latter increases the risks of disunity and even of collisions. Similarly, which better achieves Lanchestrian concentration, the entire fleet manoeuvring in a single line, or separating into independent divisions of four ships? Divisional tactics might enable more ships to take advantage of fleeting opportunities for engagement, but a single line will, if the opportunity arises, enable the full fleet to engage and gain the Lanchestrian advantage of its numbers.

Lanchestrian thinking determines not only how to handle a fleet but also what fleet to build in the first place. A crucial figure linking the two is John Jellicoe, usually remembered as the British Admiral at Jutland, but earlier Director of Naval Gunnery and one of the architects of the fleet he would later lead into battle. He was present during many of the critical planning meetings in the period 1905–1910 and here his contemplative, technocratic characteristics were of crucial importance. Of special importance was what has since become known as the "Fusion Committee" of 1905–1906, which Fisher instigated while *Dreadnought* was still not yet in commission, and which he intended should decide on another "plunging" innovation, of building the best possible super-dreadnoughts immediately.

An important conclusion from Lanchestrian thought is the pre-eminence of volume of fire: that a sufficient margin of advantage in weight of fire will dominate other disadvantages and ensure a victory if the geometry of the battle follows a Lanchestrian pattern, *i.e.* all ships are able to engage the enemy simultaneously. The Fusion Committee, with Jellicoe on it, therefore decided that the British Grand Fleet (GF) must mount more big guns even if they were on ships of lesser individual quality. Jellicoe thereby determined with the nature of the British fleet the kind of battle he must fight: the British battle line at Jutland was a dominating weapon, but was determined at least as much by Jellicoe the committee member as by Jellicoe the battle commander. Fiske's explicit identification of the crucial first five minutes also suggests the importance of range: having guns that can reach the enemy before his can reach you is a crucial advantage. "More big guns" thus give a dual advantage: of weight of shell and of longer range.

Finally, a more subtle point emerges, which turns out to be of more than merely mathematical importance. If attrition is the goal during naval combat, and during any initial period of dominance especially, this is attrition of – what exactly? Ships? Guns? The question is more complex than it sounds. Essentially the Lanchestrian unit is that which stands and falls as a single entity. The larger the unit that is degraded the more random the combat becomes, and the more unstable is a seemingly balanced fight – the outcome could go either way. If a single hit can destroy an entire ship then there is a greater risk that even a greatly numerically superior initial force can be dangerously depleted by a few lucky hits early in a combat. In mathematical language, we would say that this model is noisier, more random, more stochastic. If on the other hand the depletion is of, say, guns, of which there are more on the ship, or perhaps some even finer unit (armour quality, say, if armour can be continuously degraded and survive only a finite number of hits) then the opposite is true: the combat becomes more predictable, more continuous, more deterministic. In this case, a weaker force has far less chance of fortuitously swinging a battle in its favour. This is still a point of discussion, but echoing Fiske we have taken the turret as the optimal unit of attrition: it is typically the unit which stands and falls together.

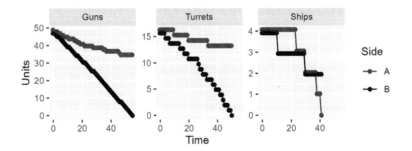

FIGURE 2.2 Examples of a simple stochastic Lanchester model. In this simplified example opposing squadrons, notated "A" and "B", of four 12-gun ships clash. Nominally A is the expected victor. In the first panel, we assume guns to be the offensive unit and in this case, the curves are slow and gradual and are very similar to the expected result from a smooth non-random simulation. In the second panel, exactly the same parameters are used but now turrets, with three guns each, are the unit which is lost on a successful enemy hit. The result is noisier, but the outcome is unchanged. In the final panel, we treat each ship as the unit (of 12 guns each). Now the simulation proceeds with sharp changes and the chance of success for "B" is increased by virtue of the disproportionate impact of a lucky shot.

There are many instances of a turret being rendered inoperable by shell hits during the First World War, but almost never a single gun.[12] But this choice is clearly vitiated if a single shot can destroy a ship – as shown in Figure 2.2 – and as happened to British battlecruisers at Jutland.

To summarize, if we believe that all-big-gun naval combat is Lanchestrian, then the goal is to create a fleet which delivers (1) a high weight of shell, (2) from a superior range and (3) in a coordinated fashion. Furthermore, (4) accuracy should be sufficient and, more subtly, especially if our fleet is superior, it is important that the potential for randomness be reduced by (5) making ships able to sustain damage that does not rapidly degrade their ability to retaliate. It is striking that, in the interplay between the technological enthusiasm of Admiral Fisher and the technocratic caution of Jellicoe and others, the Royal Navy's shipbuilding decisions resulted in a fleet which clearly met both (1) and (2). In terms of weight of shell, although the penetrating power of British shells was questionable despite the high power of the guns they were fired from, the enormous kinetic and explosive energy of shell hits on dreadnoughts affected delicate targeting instrumentation and shook crews even if they did not cause debilitating damage. Furthermore, the fact that Jellicoe, once appointed as commander of the GF, immediately prioritized points (3) and (4) is suggestive of a well-conceived plan.

However, it would be misleading to speak of the GF as a single unit. The commander of the battlecruisers, David Beatty, was less committed to the briefing and training which (3) and (4) necessitate. This mattered, because the battlecruisers fought detached from the GF as its scouting force before joining the line of battle when this role was accomplished.

Finally (5), more subtle in character, was under-appreciated at the time and this lack of understanding was compounded by Fisher's constant drive to create an HMS *Unapproachable*, his notional supremely fast battlecruiser able to outrun and outgun any opponent. But HMS *Unapproachable* is not HMS *Impervious*.

BATTLECRUISERS IN ACTION

The first real battlecruiser emerged as HMS *Invincible* in 1907.[13] Able to steam at over 25 knots and mounting eight 12″ guns, the *Invincibles* were the first in a series of battlecruiser classes – the *Indefatigables*, then the famous "splendid cats" (HMS *Lion* and HMS *Tiger*, with HMS *Princess Royal*), HMS *Queen Mary*, the two *Renowns* and finally HMS *Hood* – that saw progressive upgrading of speed, armour and gun calibre. In order not to add needlessly to the reams of writing on these famous ships, we will restrict our commentary to just the five criteria above. As super-cruisers, the ships clearly fulfilled a role in projecting British power around the globe, and they were able to outgun or outrun any competitor in far-distant oceans. However, the pertinent issue was how they would fare against enemy battlecruisers or when, their scouting duties completed, they faced dangerous foes as part of the battle line – for Britain needed them to complete its required margin of superiority over the German fleets.

The first test of these new ships came during the early months of the war. The German navy began the war with many cruisers and smaller ships spread throughout the world. These would be deployed as commerce raiders, able to harass British, French and Russian assets around the globe and thus force Britain to break up its Home Fleet and deploy disproportionate numbers far abroad to catch these nuisance ships. The most famous instance was SMS *Emden* and her activities in the Indian Ocean, but the remainder of the German East Asia fleet, including *Emden*'s sister ship SMS *Dresden* and the two 8″ gun armoured cruisers *Gneisenau* and *Scharnhorst*, formed a more powerful squadron that sailed east, rather than west, across the Pacific in an attempt to return to Germany. The squadron destroyed a weak British force at Coronel in Chile, forcing the British to despatch a strong force formed around two *Invincible*-class battlecruisers, HMS *Invincible* herself and HMS *Inflexible*.

This was the first test in anger of Fisher's battlecruisers and in many ways they performed exactly the role expected of them, and with aplomb. Outgunning the German cruisers by a significant margin and able to run them down as they fled from Port Stanley, the two superior ships nonetheless received some 40 hits from the 8″ shells of *Gneisenau* and *Scharnhorst* before the battle concluded. Although casualties were slight, damage in *Invincible* was considerable, and to have received such a large number of hits in a one-sided combat should have provoked more concern among the British, not least when one 8″ shell was discovered to have come alarmingly close to a main-battery magazine in *Invincible*. Whilst seemingly an innocuous detail, this event is a painful herald of the weakness of British battlecruisers under our fifth criterion: if a chance event can remove a whole ship at once, typically with four turrets, then this increases the randomness of the battle and the potential for a weaker force to prevail against a stronger one.

It is striking, given the hunger in naval circles for inferring conclusions from single encounters, that this encounter involving its own ships was not more rigorously studied in the Royal Navy. Perhaps the personal dislike that existed between Fisher and the commanding officer, Doveton Sturdee, did not help.[14] Nevertheless, the failure to contemplate the performance of the Royal Navy in victory is a theme we will return to. The British public expected overwhelming victories. As long as they received them, it appears the Admiralty were content, perhaps even complacent.

The two principal ships returned from the South Atlantic after repairs in Gibraltar late in February 1915, and in doing so, they missed the action to which we will devote the next few pages, the Battle of the Dogger Bank. On land, the war of movement had concluded at the Battle of the Marne, and, as the two sides settled in for the grievous war of attrition that was to follow, the war at sea began to take shape.

THE WAR AT SEA DEVELOPS

From the German Bight, the island of Great Britain appears as an enormous breakwater confining German maritime ambition. The British took advantage of this, abandoning the close blockade of the enemy coast of 100 years before, and instead imposing a more distant blockade. The main weapon was the GF, the central Lanchestrian fleet of 25–30 battleships, based at Scapa Flow in the Orkney Islands. Meanwhile the 10–12 faster ships, somewhat inappropriately termed the Battle Cruiser Fleet (BCF), were based separately at Rosyth, near Edinburgh, and enjoyed a quasi-autonomous role in scouting and harassing the Germans in the North Sea. The Battle of Heligoland Bight was a brief encounter where the battlecruisers operated independently of the GF to support lighter vessels.

Across the North Sea, the Germans had a force that mirrored the British. The High Seas Fleet (HSF) consisted of both modern dreadnought battleships and, to increase the numbers, pre-dreadnought ships, together with a faster group of battlecruisers called the Scouting Group (SG). The German force was numerically inferior to the British, but together the two groups of ships were a significant danger to either of the British forces. German Grand Admiral Alfred von Tirpitz had devised a strategy based on a German:British ratio of 2:3, so that the German fleet could be risked in attempting to cripple at least a significant portion of the British fleet, which would then be unable to defend British trade and Empire. Unfortunately for Tirpitz, British diplomacy, combined with the destruction of the Russian fleet and the clashing and restrictive orders of the Kaiser, prevented the fleet from ever being deployed in the way he anticipated. Indeed, early in the conflict, in late 1914, the then Admiral of the HSF, von Ingenohl, arguably backed away from a possible opportunity to risk his *Risikoflotte* in the way Tirpitz had intended. The SG's provocative shellings of Scarborough, Whitby and Hartlepool, which earned its commander von Hipper the nickname "baby-killer" in the British press, were of course attempts to draw the British fleet out – or, rather, to use British public opinion to force the Royal Navy to send out small groups which could then be defeated by the Germans in detail. The movement of the battlecruisers south to Rosyth in order to respond more quickly to such attacks was a partial success for this strategy.

Franz Hipper, the commander of von Ingenohl's SG, was an accomplished and competent cruiser commander, and an efficient and punctilious admiral. In the gap between Dogger Bank and Jutland, Hipper was to spend a considerable time on sick leave with combat fatigue. His opposite number was Admiral David Beatty, the commander of the British BCF and a controversial character of huge importance in our discussion. From minor aristocratic stock, he made his way into the Royal Navy via a standard route. From there, his bravery in command of various enterprises led to rapid promotion, helped by his financial independence, his exposure to active battle service and connections with crucial members of high society, including Winston Churchill, then First Lord of the Admiralty.

Beatty was an impulsive commander with considerable *élan* and charisma. He was also very aggressive and was no doubt delighted by some of the favourable comparisons to Nelson he received. As a British aristocrat, he was also a keen hunter and was guilty on more than one occasion of class prejudice, notably in his appointment of Ralph Seymour as his signals officer. He thus contrasted greatly with his immediate superior in the home fleet, Admiral John Jellicoe, who was from a far less advantaged background. Close to the outbreak of the war Beatty was appointed commander first of a battlecruiser squadron, and then of the entire BCF. Almost immediately Beatty drafted new standing orders emphasizing the independence of ships' captains in the Nelsonian tradition, in the manner of Tryon, but in contrast to the centralized control that dominated Jellicoe's GF Battle Orders, so important to the fleet as a Lanchestrian force. Beatty, as First Sea Lord between the world wars, became a crucial figure in the development of the Royal Navy into the aggressive and outstanding service of the Second World War. But his and Jellicoe's legacies were always seen principally through the multifaceted lens of perspectives on the First World War, and Jutland in particular.

The first major battlecruiser action came when, late in January 1915, with the weather changing from mild and unsettled to cold and snowy, Beatty set out from Rosyth to investigate reports that the German SG had once more set out into the North Sea. There would be no repeat of poor communication between groups or, worse still, the failure to exploit effectively the decrypted German signalling information emerging from the Admiralty's Room 40 that led to the humiliating shelling of British towns in the previous month. Beatty led a powerful force. From the 1st Battle Cruiser Squadron (1BCS), he had three of the "cats": his flagship HMS

Lion, the curiously manned *Tiger*,[15] and *Princess Royal*, although the crack gunnery of *Queen Mary* was absent owing to a refit. But 2BCS was in a state of flux – *Invincible* was under repair after the Falklands, *Inflexible* and *Indefatigable* were exchanging roles in the Mediterranean, and HMAS *Australia* was *en route* from Gibraltar after work in the South Seas, so that only *Indomitable* and *New Zealand* were available. The speed difference of these earlier ships relative to the "cats" would be an issue. Altogether this meant that out of the potentially overwhelming force of ten battle cruisers the BCF would only be able to deploy five.

A comprehensive, contemporary diary account is given by midshipman Thomas Elmhirst of the *Princess Royal*.[16] On the 23rd of January, he recounts:

> At 5.30 pm orders are received to raise steam & we left Rosyth at 6.30 pm on Saturday in XXX with *Lion*, *Tiger*, *New Zealand* and *Indomitable*. The *Queen Mary* having left for Portsmouth a few days ago to refit.

By the morning:

> The Green watch had the middle & were relieved at 6.30 am in the morning. The search lights crew started striking down the 4 aft searchlights at once. While they were doing so a lot of flares were seen off the port bow that showed that our light cruisers were in touch with the enemy. Action stations were now sounded off & half an hour later we sighted the enemy.

Who were the enemies? Hipper had four ships to Beatty's five. Three of the ships were first rate, SMS *Seydlitz*, *Moltke* and *Derfflinger*. All were slightly different. *Seydlitz* and its slightly inferior predecessor *Moltke* had five twin turrets each, with a pair of super-firing turrets (one over the other with a shared magazine) as well as the PQ wing turret configuration also present on the British cats and subsequently abandoned by both navies. *Derfflinger*, with its four turrets all on the centre line fore and aft in the manner of the British *Queen Elizabeth* class battleships, was more recognizable in the ultimate battleship configuration and was the most powerful German battlecruiser at the time. The fourth active battlecruiser, and the oldest, *Von der Tann*, was being refitted and was replaced by her predecessor, the dangerously outclassed armoured cruiser *Blücher*.

The ships moved into battle lines (in the orders given above) and began firing at their equivalent numbers in the opposing line. The British cats had a significant range advantage but were insufficiently accurate to gain the vital early advantage in hits. Further, after nearly 20 minutes of firing at ranges and speeds completely novel in naval combat, Beatty intentionally closed the range, or permitted it to close. British fire distribution was poor, with *Seydlitz* and *Blücher* attracting almost all British fire, and the combat was conducted at speeds such that the slower ships were unable to join the combat. Almost an hour into the action a shell from *Lion* found its mark on *Seydlitz*, hitting the superfiring turret, and almost caused a potentially terminal magazine explosion. But soon this success, one of only six hits on the first three German ships, was followed by repeated hits on the British ships.

Lion took the brunt of the incoming fire. She first started taking in water and began listing and later took a near-fatal hit on her forward turret that could easily have resulted in a magazine explosion. She was forced to retire from the action having taken what was later counted as 17 hits, including one that cut off her electrical power. This resulted, as flagship, in a number of confusing signals by Seymour, which ultimately led to the truncation of the battle. *Tiger* took seven hits, with only light damage. On both ships, the loss of life was relatively slight, with 12 killed and 20 wounded. On the German side, the single hit on *Seydlitz* had caused nearly 200 casualties but worse was to come when a shell from *Princess Royal* struck the lagging *Blücher*. Her engine power declined and she was left to her fate by the three leading ships. She was pounded by over 50 shells from the remaining British ships, then torpedoed and sunk. Around 80% of her 1,000 crew were killed.

This crude scoresheet seems to demonstrate an unambiguous British win – no ships lost, casualties slight, a spectacular hit on a German ship and another German ship destroyed. Of course, *Lion* had to limp back to Rosyth escorted by *Indomitable*, but she was repaired to fight again. Elmhirst:

> Congratulations are being showered on the squadron from all sides, conspicuous amongst these being some from St. Petersburg Petrograd[17] & the Russian Admiral von Essen.

And this was reflected in the press of the time and even amongst the public in the Orkneys,[18] the GF's base:

> The German Battle Cruiser *Blucher* [sic] being sunk & two more German ships being seriously damaged[.]

The Royal Navy was not to be denied its victory; this was what the public expected. But 16 months later when the entirety of the two fleets met at Jutland in an indecisive clash that is still discussed and debated today, the public would be far less sympathetic.

JUTLAND: A FIRST LOOK

The Battle of Jutland abounds in ambiguities. Fought on the 31st of May and the small hours of the 1 June 1916, and the only clash of the full British and German battle fleets, the battle concluded when the German HSF disengaged at the second attempt and reached port largely intact, ahead of the pursuing GF. The Germans were thus first to transmit news of the battle and immediately claimed victory (see Figure 2.3 as an example), having inflicted considerably more human and material loss than they had suffered, and

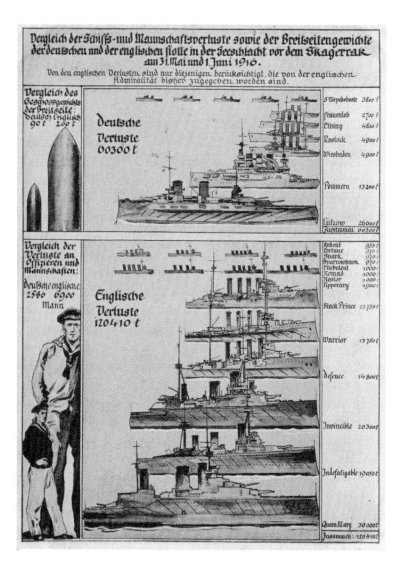

FIGURE 2.3   With the German ships arriving back in their home ports considerably before the British, the Germans were able to mount an effective information campaign which further complicated perceptions of the uncertain outcome. This image is typical of the kind of information that was released. (German Imperial Navy)

in the process apparently demonstrated technological and tactical superiority. Nevertheless, they had been forced from the field and narrowly escaped annihilation at the hands of their larger and more powerful opponent. Their original objective, to equalise this power imbalance by isolating and destroying a portion of the GF, had also failed. Success would have transformed the strategic realities of the War. As an American newspaper put it, "the German fleet had assaulted its jailor but was still in jail".[19]

The Battle of Jutland had every prospect of being the great Lanchestrian clash of dreadnoughts – to the Germans, *Der Tag*. The full German fleet, of dreadnought and pre-dreadnought battleships under Scheer and battlecruisers under Hipper, sallied into the North Sea. The British, informed by decrypts, countered with their full force of dreadnought battleships from Scapa Flow under Jellicoe and battlecruisers from Rosyth under Beatty. The German battlecruisers encountered Beatty, and led him towards Scheer in a clash of battlecruisers, the "Run to the South", which, visibility conditions excepted, followed an almost identical pattern to Dogger Bank. The outcome was very different, and fundamentally altered the later course of the battle. To which side's benefit is unclear,[20] but the loss of life and ships was painful and brutal for the Royal Navy and for Britain.[21] At Jutland Beatty once more commanded the BCF, and had the six battle cruisers of the 1BCS and 2BCS – the latter now consisting only of *Indefatigable* and *New Zealand* as HMAS *Australia* was absent due to a collision – and four *Queen Elizabeth* class fast battleships of the 5th Battle Squadron (5BS), regarded as the most powerful ships afloat. Armed with eight 15″ guns and capable of 23 knots, these ships were temporarily attached to the BCF in place of the three *Invincibles* of 3BCS which were, somewhat belatedly, undertaking gunnery training at Scapa Flow with the Battle Fleet. When the Run to the South commenced, a running battle took place between the six ships of 1 and 2BCS and the five ships of the SG – SMS *von der Tann* at the rear, and *Lützow*, sister ship to *Derfflinger*, in the van joining the three surviving veterans of Dogger Bank.

The British were to lose two ships (*Indefatigable* and *Queen Mary*) to magazine explosions and were fortunate not to lose their flagship *Lion* to another. The Germans, until accurate and punishing 15″ fire from 5BS began, were relatively unscathed. Beatty was famously heard to remark "There seems to be something wrong with our bloody ships today".

WAS THERE SOMETHING WRONG WITH THEIR "BLOODY SHIPS"?

So it looks as if the British were lucky at Dogger Bank and then unlucky at Jutland. Could it have been otherwise? Of course, had the British been unlucky at Dogger Bank, they might, like the Germans, have avoided the classic victor's error of failing to learn the battle's lessons and thereby perhaps avoided the Jutland sinkings. This is the basic premise about which we construct our question to be addressed. Beatty's failure to act on the warnings that Dogger Bank provided allows us to combine data from Dogger Bank and Jutland to estimate the parameters for a basic Lanchestrian model – rate, accuracy, damage and so on. We then challenge the model with the actual events of Dogger Bank. If they don't match up, there are essentially two possibilities: either the parameters are very far wrong – unlikely given that they combine Dogger Bank and Jutland data – or the British were lucky.[22]

We want the model itself to be stochastic (*i.e.* random): crudely, when a ship fires, it must roll dice to see whether it successfully hits its target. This is simple enough, but we have a problem compared to a typical case from experimental science: we only have one observation, that of history. In science, there are many different observations – the experiment can be repeated, often many times. With many observations, we have many numbers to fit to, and we can understand the predictability and variability of the result. With history, we do not have that luxury.

Fortunately, rapid advances in statistical computing have opened up new avenues to ask, and answer, these kinds of questions. One technique is *Approximate Bayesian Computation (ABC)*. This startlingly new approach has had many applications, from scientific areas such as ecology and evolution[23] to commercial ones such as online betting. Suppose we have a model of a process and a set of parameters which we think we can feed into the model to describe some data we see. Crucially, we must now adopt a Bayesian approach: rather than having a precise estimate of each parameter, we posit the ("prior") distribution of what we think are its likely values. If we have good reason to think that we know a particular parameter quite precisely then the distribution will be sharply peaked, whereas if we are uncertain then the distribution will be broad – spread out – reflecting our lack of knowledge.[24] We can now run a simulation by picking a set of values randomly from these prior distributions and using them in our model. We get an outcome. We now ask, does it match the (one) historical outcome well or badly? If it is close, we keep this trial and its set of parameter values, and stay "near" it; if not, we reject and try again. In this way, we wander in the space of all possible parameter values, trialling many and accepting only those which are both compatible with our prior beliefs and give outcomes which are close in some sense to the historical observations.

This is not dissimilar to military wargaming. We choose a set of rules which we think describe reality, we play the game, and then we assess and analyse the results. Such exercises are used routinely in the armed services, beginning with the Prussian *Kriegsspiel*, intensifying between the wars with huge games played out in naval tactical schools in Britain and the US, and continuing to this day. However, there is a crucial difference. Between the wars, Jutland was gamed with one set of rules perhaps once a year. But with modern computing, several *billion* instances may be processed.[25] This level of exploration allows statistical comparisons of results and the ability to say with some confidence how findings compare with previous beliefs. Such comparisons are the hallmark of Bayesian thought, which gives not absolute results

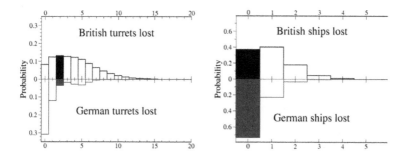

FIGURE 2.4   Losses of turrets and ships at Dogger Bank from the ABC simulation – figure reworked from that appearing in MacKay, Price and Wood, 2015. The solid-filled bars indicate the actual result – noting that the simulation terminates at the point of disengagement so *Blücher* is still afloat and firing from all six turrets.

but rather tells us how our multiple computerized wargames challenge our beliefs with historical reality and should modify them.

So what was found? First and foremost the results at Dogger Bank appear to have been very unlikely, see Figure 2.4. At Jutland, with a much larger sample, only the most recent of the British battlecruisers, HMS *Tiger* was able to take as much incoming fire as did *Lion* at Dogger Bank without catastrophic results. By comparison, with only a small number of shots the British were able at Dogger Bank to inflict an almost-fatal hit on *Seydlitz* which destroyed two turrets, when no comparable such events happened, or came close to happening (at least in the early phases) at Jutland.

The eventual destruction of *Blücher* distorted the British view of Dogger Bank, facilitating a narrative of expected victory which suited Admiralty and British public alike, so that no one analysed the performance of the battlecruisers in the earlier action that really mattered, the running battle between comparable ships. But the signs of trouble were there, observed by junior officers: for example, Lt Henry Blagrove wrote that "We were marvellously lucky to escape as we did as their shooting was damned good",[26] while Elmhirst said that the Germans "straddled us repeatedly, ...

all their shots falling and guns firing extremely well together".[27] The British ships behaved at Jutland exactly as could be reasonably expected. If anything the problem lay in their not being blooded enough: Dogger Bank was not so much a lost opportunity for British victory as a lost opportunity to pursue analysis that might have led to very different outcomes later. Our simulations show that the most probable results at Dogger Bank would have been for the British to lose a ship or two, while the Germans – although the vulnerable *Blücher* was still a liability and a possible loss – were perhaps unlucky. Of course, as with all counterfactuals, we must avoid pure flights of fancy, and events might well have developed very differently once ships had begun to be seriously damaged in ways different from what actually transpired.

REFIGHTING JUTLAND?

The combination of Dogger Bank and Jutland allows us to establish the basic model very strongly. However, the model on its own is not enough to fight counterfactual battles of Jutland, which would be crucially dependent on the fleet geometry that develops, with randomness and commanders' decisions, some of them in response to each other, allowing wild ramifications. But the Jutland battlecruiser action, beginning with the Run to the South, is, like Dogger Bank, mere calculus, a relatively simple engagement between two battle lines, and this we *can* now reliably simulate as long as we do not diverge too much geometrically from what happened. Moreover, one of the most crucial questions of Jutland is what would have happened if the 5BS (of *Queen Elizabeth*-class super-dreadnoughts, which we recall had been detached from Jellicoe's command to join Beatty's battlecruisers while 3BCS was training at Scapa Flow), had been properly integrated into Beatty's battle line. And this question we can answer, or at least address.

The 5BS was the most powerful unit of ships in the world in 1916 and remained a formidable force throughout their lifetimes (see Figure 2.5). One of their number, HMS *Warspite*, would fight through the Second World War to become the "Grand Old Lady"

FIGURE 2.5 The ships of the 5BS led by HMS *Barham*, photographed here early in the Second World War, post modernization. The ships played a key role in both wars, but *Barham* was lost to a U-Boat attack in the Mediterranean not long after this photo was taken. (Daily Mail/Shutterstock)

of the fleet, with the most battle honours of any British warship. It would be a reasonable expectation that any admiral fortunate enough to command such a squadron would expend considerable thought on how they might best be deployed, not least because they were at least three knots slower than the rest of the BCF, though not outpaced by the German Battlecruisers in action. The commander in charge of the 5BS was Hugh Evan-Thomas, an able stalwart of Jellicoe's command. Yet he was barely briefed by Beatty upon his arrival at Rosyth, three days prior to the ships' departure for the Battle at Jutland. There is apparently no record of any meetings.[28]

Let us revisit and review what actually occurred. Jutland was one of the largest naval battles ever fought, and probably the most controversial. It was initiated when the British responded to intelligence that the German fleet was putting to sea. Their plan was simple: lure out the British battlecruisers, trap them – preferably between the two German groups – and destroy them. In Scapa Flow, the Battle Fleet, with Jellicoe in command, slipped out to attempt a plan which mirrored the Germans'. The Battle Fleet, although slower than its battlecruisers, was a carefully balanced Lanchestrian force, built to maximize the weight of fire, drilled to maximize central control and meticulously practised to maximize rate and accuracy of fire. It was unquestionably the fleet that Jellicoe had conceived, but would the battlecruisers succeed in enabling its deployment in the manner intended?

The event itself was as murderous as it was inconclusive. The opening action between the BCF and its opposite number the SG, the Run to the South, was almost ruinous for the British and led to the loss of two of its ten battle cruisers. Beatty's flagship only escaped joining them owing to the extraordinary bravery and sacrifice of the commander of the marines in charge of one of his turrets.[29] This phase of the battle came to a close as Scheer's HSF arrived and the British ships reversed course, fighting a running battle with both Scheer's and Hipper's ships, the "Run to the North". Whereas the 5BS was only able to force a minor role in the southwards battle, it was very much to the fore in the "Run to the North". Although these powerful ships endured a torrid moment as they turned[30] into the teeth of Scheer's onrushing ships, it seems reasonable[31] to assume that Evan-Thomas also ran a dangerous course partly to protect and shield Beatty's damaged BCF with his better-armoured ships and 15″ gunnery.

For the Germans, their trap had almost sprung, but the mouse had hopped out just as the bar came down. They pursued relentlessly, but in doing so nearly inverted the situation, for they ran straight into Jellicoe, who with poor information and little time had nevertheless utilized his effective central command to deploy his two dozen battleships into a single line, an arc awaiting Scheer's arrival from the south-west. There are many details here crucial to the final outcome but not to our discussion. *Warspite* endured a risky moment as her steering jammed and she performed a circular dance at "Windy Corner"; the battlecruisers of Admiral Hood's 3BCS caused considerable confusion to the Germans as they ranged ahead of Jellicoe[32] but were to lose *Invincible* and Hood himself; many British casualties were caused by outdated lighter ships attempting to settle scores. The confusion was increased by Beatty steaming across the front of the entire battle fleet to get to the front of the line; Evan-Thomas joined its rear.

What followed next was a confused action in fading light, which many commentators perceive as a failure of Jellicoe to press home his advantage. Others perceive that his primary goal had to be to preserve his Lanchestrian fleet. With the benefit of our analysis of Dogger, we can reasonably suggest that some sort of general pursuit – realistically the only way Jellicoe could have prosecuted the battle more aggressively – could have dangerously increased the randomness of the action. For a superior Lanchestrian fleet, this could not have been an appropriate step. In any case, the Germans were twice able to execute 180-degree battle turns in the teeth of the GF, the first rather more orderly than the second, covered by Hipper, several of whose ships were barely able to limp back to port. *Lützow*, his flagship and the

FIGURE 2.6 SMS *Seydlitz* on her return to Wilhelmshaven after the Battle of Jutland. Her wrecked but still floating state is a commentary on both the build quality of the German ships and the amount of damage received from the British. It does not seem controversial to suggest that a British ship could not have sustained this level of damage and stayed afloat with her crew, which partly explains the disparity in the loss of life between the two sides. (German government)

pre-dreadnought *Pommern* were lost, while *Seydlitz*, *von der Tann* and *Derfflinger* were little more than floating wrecks on their return (Figure 2.6), their survival a testament to their builders and their close proximity to their base. The German fleet then escaped under cover of night back to Wilhelmshaven. The dreadnoughts and battleships of the HSF were never to put sea again in anger, their last act instead being to surrender in Scapa Flow before being scuttled.[33]

For the British, this was an effective strategic victory, but a tactical defeat in material and to the media. The news of the battle was poorly handled and the imbalance in the loss of life and ships was clear. The actions, and perhaps mis-actions, of the admirals have been debated at length for the last hundred years. Yet the outcome, that the powerful German surface fleet would play no further role in the outcome of the war, can have been no more than what could have reasonably been expected. Different strategic outcomes are the realms of the exuberant counterfactual. Could a victorious German fleet have severed trade and cross-channel communication and brought Britain to its knees? Or could its complete destruction have precipitated a successful amphibious landing in Pomerania, shortening the war? These are open counterfactual questions beyond our methods.

TO FOLLOW OR NOT TO FOLLOW?

But let us return to the beginning of the action of Jutland, and a moment of considerable controversy that is worth dwelling on. As was standard submarine-avoidance practice, all ships zig-zagged while at sea. As the 5BS was tracking to the north-east, Beatty received the report from HMS *Galatea* of contact with German forces and ordered the BCF to turn south-south-east. This report was not relayed to the 5BS by HMS *Tiger*, and no *explicit* orders were given to 5BS – Beatty's standing orders for the BCF meant that there was an *implicit* expectation for Evan-Thomas to follow Beatty's lead in *Lion*. But of course, Evan-Thomas had not been briefed by Beatty, and the standing orders of the GF, with the goal of Lanchestrian concentration to the fore, would not support this interpretation.

In constructing suitable restrained counterfactuals, this crucial moment is our starting point. It led to the loss of five minutes of engagement – Fiske and Baudry's crucial five minutes for a Lanchestrian engagement. As BCF commander, with the knowledge that your best ships are slower and the enemy is likely to be engaged to the south, how do you best deploy them? To the south, so that your faster ships can catch them and create a joint sledgehammer blow? Or to the north, so that they can position themselves to be an auxiliary but decisive support group, perhaps even the jaws of a baited trap? Beatty deployed them to the north, so what was Evan-Thomas to think? His subsequent decision-making during the "Run to the North", and especially during the events at its beginning and end, suggests a brave and capable commander able to make good calls when under extreme pressure.

Let us consider, then, alternative fleet geometries and policies for the Run to the South. All will be contests between Beatty's forces (BCF and 5BS) and Hipper's SG, and we assume a simple battle-line throughout. We assume no change in posture from Hipper – a reasonable, minimal assumption. We simply take a set of priors that constitute the combined knowledge of events of Jutland and Dogger bank, integrated from the data in Campbell[34] and the inferences from Okun[35] and our own work, and pose a series of differing challenges and alternative scenarios using the ten British ships that sail from Rosyth together and the five battlecruisers of the German SG. We consider the ships of 5BS to be markedly superior to the other British ships, for their accuracy is greater – they had been training alongside the Battle Fleet and contemporary accounts indicate their good shooting. Their armour is thicker and, most importantly, their vulnerability to flash explosions is greatly reduced: the ships survived numerous salvos at Jutland and throughout subsequent action in the Second World War, without any of their number perishing in this fashion.

The scenarios for the British posture and composition include different arrival times for the 5BS (Figure 2.7), differing lengths of battle, and changes in accuracy, vulnerability to flash explosions and rate of fire. Each of these changes is motivated by a particular historical circumstance. Changing the arrival time of the 5BS is directly motivated by the criticism of Hugh Evan-Thomas's failure to respond immediately to Beatty's reaction to the sighting signal from *Galatea*. Even earlier arrival times might also indicate more extreme counterfactuals where Beatty deploys his ships differently. This is also reflected in the differing lengths of the battle which effectively indicate the degrees of success of Beatty in causing the clash between the two groups to occur further to the north and

FIGURE 2.7   A First World War photograph of two of the ships of the 5BS, Warspite and Malaya, taken from a third, Valiant. This image is reputed to have been taken at 1400, just prior to first contact with the SG, but the ships' cruising configuration is unclear. (Imperial War Museums)

thus for the SG to be increasingly further away from the support of the HSF, whose arrival effectively terminates the Run to the South. Finally, we also consider the possibility that, faced with the evidence from Dogger Bank and the Falklands, the BCF makes changes to its ammunition-handling procedures, retreating from its prioritization of rate of fire and increasing its commitment to accuracy. This could have been done simply by rotating ships for gunnery training to Scapa much earlier after the clash at Dogger. For our simulation, this results in a reduction of the flash probability (at the expense of rate of fire) and increases in accuracy – essentially, we make the British ships more like the German in their firing characteristics (see Table 2.1).

First, we must compare our simulations to reality. Our baseline simulation compares our model in non-rigorous way with the actual result during the Run to the South. We find a good fit to the real data, although the simulation results suggest that the British were a little unlucky to lose two ships – a small difference, around a single standard deviation (roughly half a ship), and

TABLE 2.1 Labelling key for 5th battle squadron counterfactual scenarios. Scenario 0 is our baseline with A1, A2 and A3 then representing modified arrival times and A4 and A5 indicating lengthening battle lengths due to clash occurring further from the High Seas Fleet

Scenario	5th Battle Squadron Arrival Time	Length of Battle (Mins)
Baseline	27	66
A1	20	66
A2	10	66
A3	0	66
A4	0	99
A5	0	132

certainly nothing like their contrasting level of positive "luckiness" at Dogger Bank. In what follows we will state deviations from this baseline simulation rather than deviations from reality, though the two are sufficiently close that this shift does not overly distort interpretation (see Table 2.2).

The first question we ask is how much the late arrival of the 5BS, by 27 minutes, affects the outcome of the battle. Perhaps surprisingly, the answer to this alone is "not much". We find that if the 5BS is not late at all, instead beginning the action simultaneously as part of a 10-ship line, then this shifts the mean number of hits on German ships up from 18 to 24, destroying one extra German turret on average, and reduces turret hits on British ships

TABLE 2.2 Mean hits sustained, turrets lost and ships sunk for British (B) and German (G). Whilst the mean number of ships sunk in our baseline simulation is only 1, the peak is quite broad, reflecting the flash probability and subsequent impact on outcome due to the drop off in shells fired

Scenario	B Hits Received	G Hits Received	B Turrets Lost	G Turrets Lost	B Ships Sunk	G Ships Sunk
Baseline	43.8	18.1	6.34	2.37	1.01	0.57
A1	43.3	19.7	6.30	2.61	0.99	0.63
A2	42.3	21.9	6.14	2.94	0.97	0.71
A3	41.4	23.7	6.03	3.19	0.95	0.77
A4	58.3	34.5	8.27	4.61	1.37	1.17
A5	72.9	44.4	10.09	5.88	1.75	1.55

downwards from just less than 6.5 to slightly over 6, again on average. Compared to the typical random variation, this is only a small improvement for the British.

Next, we ask how extending the battle alters the expected results. This reflects improved deployment and tactics from Beatty resulting in the isolation of the SG from the HSF for longer times, with the 5BS playing a full part in the action. Again the result is poor for the British, who receive more damage (in terms of turrets, the chosen Lanchestrian damage unit) than do the Germans. The problem here for the British is that simply extending the battle and thus exposure to German fire, with weak British BCs and their gunnery, results in higher casualties than the Germans'. Even with better tactics and the presence of the best ships available the expected number of ship losses is still higher for the British than for the Germans. If taken to conclusion, the most extreme example is where we imagine Admiral Scheer becomes aware of Jellicoe's presence and opts to leave the SG to its fate in a fight to the finish with the BCF. This is highly implausible, and impossibly restricted by ammunition, but it is still revealing that the SG is *victorious* – now this is a fight to the death meaning the complete destruction of the opposing side – one sixth of the time. Worse still, in over 40% of British victories, they lose five or more ships. These statistics are not encouraging for a navy attempting to assert its dominance.

For better outcomes for the British, we must consider counterfactual gunnery procedures and training (see Table 2.3). For example, suppose the British had improved their shipboard policies to reduce the probability of ruinous flash explosions, which

TABLE 2.3　Labelling of the scenarios for different training regimens of the BCF

Scenario	Accuracy Increased	Flash Fire Decreased	Rate of Fire Decreased
Baseline			
B1		✓	✓
B2	✓		
B3	✓	✓	✓

in turn necessitates a decreased rate of fire. There is considerable uncertainty here, as the high chance of flash fire on a British ship is a complicated blend of both poor ammunition practice and vulnerability to plunging fire due to weak deck armour. The latter only became apparent after the clash at Jutland – indeed it is remarkable that 25 years later HMS *Hood* was still vulnerable (although some critical changes to her armour protection were planned but never made). If rate of fire and flash are reduced, then in a conventional Run to the South (without the addition of the 5BS) the loss of British ships goes down, but so do hits to the opposition.

If now we imagine, in addition, more battlecruiser gunnery training, then accuracy improves and hits increase. Increased accuracy and reduced rate of fire have approximately equal and opposite effects during a conflict of the length of the actual Run to the South but do result in a near halving of the expected British ship losses (see Table 2.4). This brings approximate parity to the two combatants. To put it more starkly: the British battlecruiser fleet compensated for the failure to train its gun crews adequately by gambling with the lives of its sailors.

Finally, we consider everything together. What if we allow all possible improvements for the British in both deployment tactics and shipboard procedures? Here we give the British the engagement of the 5BS in the battle line from the outset, together with safer and more accurate gunnery (see Table 2.5). It is worth noting that the bases for all of these – gunnery training, flash discipline and effective tactical briefing of Evan-Thomas – were matters

TABLE 2.4 Mean hits sustained, turrets lost and ships sunk for British (B) and German (G) for different training priorities compared to the baseline

Scenario	B Hits Received	G Hits Received	B Turrets Lost	G Turrets Lost	B Ships Sunk	G Ships Sunk
Baseline	43.8	18.1	6.34	2.37	1.01	0.57
B1	44.9	13.6	5.37	1.86	0.67	0.44
B2	41.9	26.5	6.12	3.34	0.97	0.82
B3	43.8	18.1	5.25	2.37	0.65	0.57

TABLE 2.5   Labelling of the scenarios for different training regimens and battle timings

Scenario	Accuracy Increased	Flash Fire Decreased	Rate of Fire Decreased	5BS Arrival at Start	Length 132 Min
Baseline					
C1		✓	✓	✓	
C2	✓	✓	✓	✓	
C3	✓	✓	✓	✓	✓

under Beatty's control, and in which it is generally considered that he fell short, so that this is far from being a set of unrelated counterfactual suppositions but is rather a single counterfactual, of a very different and more assiduous BCF commander. Then, and only then, do we see the Run to the South resulting in a positive net outcome for the British, in the form of a larger number of German ships sunk than British (see Table 2.6). If this clash were to run to completion, it would heavily favour the British, who indeed now have a small chance of eliminating the SG entirely, although in any plausible period of time in daylight this is no greater than 2.5%.

One last question is whether, had *Seydlitz* not suffered its explosion at Dogger Bank, German battlecruisers might have had sufficiently greater flash risk to cause probable losses at Jutland. However, this is a historical "what if?" outside the scope of our modelling, for there were too few British hits on Germans at Dogger Bank to give a sound basis on which to estimate German probability of flash fire.

TABLE 2.6   Mean hits Sustained, turrets lost and ships sunk for British (B) and German (G) for different training priorities and battle configurations compared to the baseline

Scenario	B Hits Received	G Hits Received	B Turrets Lost	G Turrets Lost	B Ships Sunk	G Ships Sunk
Baseline	43.8	18.1	6.34	2.37	1.01	0.57
C1	42.4	19.1	5.09	2.66	0.63	0.64
C2	41.4	23.6	5.00	3.19	0.61	0.78
C3	73.0	44.5	8.61	5.89	1.28	1.55

This reconstructive battling enables us to put some level of statistical insight into multiple realizations of a key phase of Jutland. The model is crude and laden with assumptions – as are all wargames – but, unlike in a wargame, our goal is simply to understand what is plausible and what is not. George Box's comment that "all models are wrong, but some models are useful"[36] must surely ring louder and truer today, when so much more is possible, than it did when he made this comment (as the Second World War raged and computers were still hidden in realms of secrecy in Santa Fé and Bletchley Park). Each individual scenario in this section is run 100,000 times – completely beyond the possibility of wargaming. History provides us with only a single realization; wargaming might provide us with tens or hundreds of samples from the error distribution around the expected result; but here we are capturing a full spectrum of possible errors.

What have we learned from this exercise? Relevant to historical analysis, the criticism of Hugh Evan-Thomas is unwarranted. His delay did not make any impact on the battle that would have fundamentally altered the result. Furthermore, if one considers that he may have believed that he was playing a part in a well-conceived trap to enhance the impact of his powerful ships then his decision was a gamble worth taking. If this moment is then put aside, his conduct, through the remainder of the action, unbriefed and hampered by poor communication with his commander throughout, was excellent, if not outstanding. Instead, for criticism, we must focus on Beatty – his poor preparation of the Rosyth-based BCF, his prioritization of rate of fire over accuracy (which works to the disadvantage of the longer-ranged British guns) and his not having briefed Evan-Thomas. Whilst the threat of U-boat interference close to Rosyth during gunnery practice was real, this must be balanced against the level of threat created by their unpreparedness. Beyond this, British ship design has to remain suspect – although, as we noted, we cannot disentangle it from poor ammunition handling.

We can perhaps conclude something about counterfactual modelling more broadly. That is, we conclude nothing – or rather, we find that individual changes of the type we have explored here were unlikely to have greatly altered the main trajectory of battle. This should be regarded as a vindication of the method of restrained counterfactuals. For a reversal of fortune, we had to put various low-level counterfactuals in superposition. Yet this was not an exuberant counterfactual, but rather a single, higher level counterfactual in the form of a very different alternative to Beatty, whose preparations needed to be better if he were to win his part of the battle of Jutland. But despite this Beatty did, in the end, what was required of him: he delivered the German fleet onto the guns of Jellicoe's battle fleet, and perhaps this ultimate success and his earlier failings were alike products of his personal dynamic.

In the case of Jellicoe, too, his personal traits bind the GF and its engagement together, for it was a fleet he gunned, helped build, fought – and for which he eventually declined an uncertain outcome in the German Bight, just as he had long since made clear he would. A counterfactual Jellicoe is far beyond the reach of modelling, but it is striking that Beatty, when he took over command of the GF from Jellicoe after Jutland, became notably more cautious in his outlook. Thus, the title question "Could the Germans have won the Battle of Jutland?" is somewhat a straw man. Our quantification of the most accessible restrained counterfactuals suggests that they would have to slide from exuberance into plain fiction to lead to meaningfully different results. It is certainly true that an overwhelming British win might have led to Fisher's Baltic scheme and an invasion of Pomerania, or an overwhelming German win to a crippling of British trade and severance of Channel supply lines. However, the reality is that only extremely unlikely results from either side would result in a significantly different outcome – in both cases, the margin required is well into the tails of any sensible counterfactual distribution. In the sense of the previous chapter, we suggest that Jutland, despite its obvious prominence, is not really a critical juncture at all. That is not to say that the sum of dreadnought engagements could not have played out differently. Dogger Bank was unlikely, and its more probable outcomes would have led to a different Jutland or Jutland-like clash. Even before that, a 21cm shell came within a whisker of *Invincible*'s magazine at the battle of the Falkland Islands, and this too would have changed the historical trajectory. However, it would not have changed the ultimate outcome at Jutland or its equivalent, or the strategic British control of the seas. Rather it was the necessity to build a battlefleet capable of defeating multiple opponents, effected through the dreadnought programme and later the Fusion Committee, that determined these. Pursuit of "alternate Jutlands" will never lead to any meaningful historical insight. Rather our restrained detailed counterfactuals push us back towards the importance of individuals, their personalities and their decision-making, including long before the fighting – a theme we shall see again in the next chapter on both sides of the Battle of Britain.

NOTES

1. Churchill, *The World Crisis* 1916–1918, part I, p112.
2. Perhaps the most likely candidate would be the roughness of the sea state during the Battle of the North Cape, but ultimately this did not obviously affect the outcome.
3. Gordon, *Rules of the Game*.
4. The title of an excellent book on this era: Rodger, *Command of the Ocean*.
5. The Battle of Hampton Roads is discussed by many authors. See for example Davis, *Duel Between the First Ironclads*.
6. The event was satirized in the untimely demise of the fictional Admiral Lord Horatio D'Ascoyne in the 1949 Ealing (British) comedy *Kind Hearts and Coronets*.
7. This imbalance may have contributed to her unusual resting manner when the wreck was discovered in 2004 – burying upright, prow embedded into the sea floor. "The admiral ordered 'turn' and the ships collided" by Nicholas Blanford, *The Times*, London, England. Thursday, 2 September 2004 Issue 68170 p20.
8. Churchill's *The World Crisis* (Pt 1, Ch 5) includes a lengthy quote which summarises Fisher's view (dated 13 February 1912) on p104.

Fisher was a, if not the, prominent character of the dreadnought era. See, for example, Massie, *Dreadnought*.

9. Baudry, *The Naval Battle*; Chase, *Mathematical Investigation*; Fiske, *American Naval Policy*; Lanchester, *Aircraft in Warfare*; Osipov, *Influence of Numerical Strength*.

10. Letter from Jellicoe to Lanchester, 15 June 1916, held as B3/18, Lanchester archive, University of Coventry.

11. Capt. R. Plunkett-Drax, "Notes on Grand Fleet battle tactics", DRAX 1/18, Churchill Archive Centre, Cambridge; see also "Grand Fleet battle tactics", 1 January 1917, in BTY/7/2, Caird Library, Greenwich.

12. The French battleship *Dunkerque* had two guns knocked out in a quadruple turret during the unfortunate combat at Mers-el-Kebir.

13. *Invincible* was termed an "armoured cruiser" at her launch and strictly only became a "battlecruiser" after an Admiralty order in 1911. The armoured cruiser *Blücher* was the Germans' response to what they believed the British were building in *Invincible*.

14. Sturdee is conspicuous by his absence from Fisher's letters, but a footnote in his collected letters (Marder, *Fear God and Dread Nought*) indicates "Fisher could never have worked with Sturdee. His dislike of him went back to 1907..." (p73). Fisher goes on to call him a "pedantic ass" in a letter to Beatty in November 1914 (p82) and accuses him of "...criminal ineptitude..." to Jellicoe shortly after (p101). Fisher seems keen to glory in the speed with which he acted in dispatching Sturdee and focussed on his failure to deal with SMS *Dresden* at the Falklands.

15. HMS *Tiger* was the newest of the "cats". With more powerful and better protected secondary armament, she was seemingly a more resilient ship than her sisters. Yet Beatty was to write "I did say that 'Tiger' had a very mixed Ship's Company, with a large number of recovered deserters, and that it was an uphill task for the Captain to pull them together in War Time, and the same efficiency could not be expected from the 'Tiger' as from the other ships." (Letter from Beatty to Second Sea Lord Sir Frederick T. Hamilton dated 17 Feb 1915, National Maritime Museum, HTN/117, reproduced on p249 of Vol 1 of The Beatty papers (ed. Ranft).

16. From a journal account of the Battle of Dogger Bank by an officer of HMS *Princess Royal*, in ELMT 1/5, Churchill Archive Centre.

17. St Petersburg became Petrograd on 1 September 1914. It is curious that this is edited in the diary.

18. Diary entry for the 29 January 1915 from Margaret Tait of Kirkwall held in the Orkney Archive, D1/525.

19. *New York Herald*, 3 June 1916, as quoted by Holger Herwig, Naval War College Review 60 no. 2 p163.
20. It can only be exuberant to speculate what would have happened had Beatty's force fought off the SG with minimal losses and thus encountered the HSF whilst still a largely intact and coherent body of ships.
21. An excellent general text on Jutland is Steel and Hart, *Jutland 1916*.
22. A third apparent possibility is beyond the scope of the model: that the Germans acted on the information they gained the hard way at Dogger Bank, through *Seydlitz's* explosion, and thereby shifted the odds in their favour for Jutland. Whilst possibly true, this cannot be tested in the simulation as too few hits were scored on the German ships by the British during the running action at Dogger Bank to give us useful information. However, it is worth noting that *Blücher's* stubborn endurance of later shelling stands in contrast to the flash explosion that occurred on *Seydlitz*. It does not necessarily follow from *Seydlitz's* explosion that German flash-fire probability at Jutland was lower than it was at Dogger Bank.
23. See, for example, Csilléry et al., "ABC in practice".
24. But the distribution is not flat! A true Bayesian accepts that they always have some prior belief, however vague, and statistical theory tells us the natural forms taken by these.
25. As, for example, in MacKay, Price and Wood, *Dogger Bank* and *Weighing the Fog of War*.
26. Lt (later Rear Adm) Henry B C Blagrove, letter to Lt Oswald Freeth, HMS *Tiger*, 6 March 1915, DRAX 1/47.
27. Blagrove and Elmhirst, cited above.
28. Discussed in Gordon, *Rules of the Game*, p54–58.
29. Major Frances Harvey was posthumously awarded the Victoria Cross for his actions in flooding Q turret magazine.
30. They did this by all making their turn in the same patch of sea, one after the other, so that their order was preserved – very dangerous when under fire, as the enemy has no need to alter its gunnery solution.
31. Evan-Thomas ordered a turn to starboard to keep himself on a parallel course to Beatty when he could easily have interpreted Beatty's signals as an order to follow Beatty's precise course.
32. Evidence from Jutland anecdotally suggests the shooting of the three *Invincibles*, which had been practising at Scapa Flow with the GF, was notably improved. Gunnery practice was not possible at Rosyth.

33. They later became a source of high-grade steel for space missions. It is a curious thought that the furthest man-made object from earth, the *Voyager* probe, contains metal from the German battle-fleet, just as the first broadcast from earth was that of Adolf Hitler. The pre-dreadnoughts had a more varied life, and the Jutland veteran *Schleswig-Holstein* was to fire the first shots of the Second World War.
34. Campbell, *Jutland*.
35. Okun, *Naval Gun and Armor*.
36. A paraphrase of the argument of Box, *Science and Statistics*.

CHAPTER 3

Could the Germans Have Won the Battle of Britain?

FIGURE 3.1 German Heinkel He 111 Bomber over east London. (Luftwaffe, photographer unknown)

DOI: 10.1201/9780429488405-3

HISTORICAL CONTEXT

In the Second World War, 24 years after Jutland, the first purely attritional air campaign was fought in the skies over southern England in the summer and autumn of 1940. Germany was attempting to gain air supremacy, either to force Britain to make peace or, if it would not do so, to enable an invasion. The attempt failed, and its repulse has become a national myth in the UK. Part of the myth is that the margin of victory was narrow – which immediately drives us to ask, "how narrow?" Could the outcome have been otherwise? If so, what different decisions would have been needed, and from whom, with what different mindsets?

In the previous chapter, we saw that the core of dreadnought naval warfare was Lanchestrian – that each side caused losses in proportion to its numbers – and we used this model to explore its outcome. But Lanchester's original intention was to describe the role of aircraft in war.[1] So is air combat best described using Lanchester's model? As we shall see, the answer is a qualified "no". Instead, we will need to use a different approach to describe historical combat which makes no assumptions at all about the nature of that combat, and which needs no underlying model, but which is instead based on the statistical idea of "bootstrapping".[2]

At its simplest, the resource controlled by the commander in an air campaign is *sorties*. He or she has a number of aircraft available of different types and must choose whether and when to fly them. For fighters, to state the obvious, the crucial thing is that sorties should tend to cause kills but not losses. Of course, attacking fighters are also there to protect bombers from losses, and defending fighters are there to prevent or disrupt the bombers' missions, but the *attritional* campaign will certainly be measured by the casualty exchange ratio, the ratio of the two sides' losses.

In Lanchester's model, this works very simply: each side's losses are proportional to its enemy's numbers. But, unlike in dreadnought naval warfare, in air war there's no simple rationale for such an assumption. It's not at all obvious that, for a given number of aircraft attacked, twice as many attackers will achieve twice as many kills, and for no additional losses. An alternative, very simple, hypothesis might be that air combat is essentially just a set of duels, very much like the notion of the battle line in Lanchester's "ancient" warfare model: the absolute numbers of aircraft engaged on each side may vary, but the ratio remains roughly constant. Whereas in the Lanchester model the casualty exchange ratio is proportional to the force ratio, in the set-of-duels model the casualty exchange ratio is independent of force numbers.

THE BATTLE OF BRITAIN

We can see how such ideas play out in the data from the Battle of Britain.[3] Germany, as it approached its final victory over France, made a concerted attempt with its air arm, the Luftwaffe, to gain air supremacy over England by destroying the British air force (the Royal Air Force, or RAF). Whether the RAF was to be destroyed in the air or on the ground, and whether an air victory alone would achieve German war aims – perhaps through negotiation or British capitulation – or rather enable an invasion, were for the Luftwaffe secondary questions. Day after day through the summer of 1940, German bombers made their attacks – first on Channel shipping and ports, later on airfields and finally on London, as in Figure 3.1 – escorted by fighters. Bombing worked in two ways: by destroying targets on the ground, and by drawing up fighters to defend them, which could then be attacked by German fighters. The RAF needed to destroy – or at least disrupt and frustrate – the bombers, whilst not allowing its fighters to become prey for the Germans. For the British, the battle was a holding action: frustrating German aims, creating delay until the short days and bad flying weather of winter, were the goal.

To achieve victory in this holding action, RAF Fighter Command needed to remain a force-in-being and to demonstrate this to Hitler. An analogy follows: fighting close to its own airfields, and with the Luftwaffe uncertain about the effects its surface attacks were having, the RAF was conducting a kind of "reverse-slope defence" such as that used by Wellington at Waterloo. Keep your forces behind the crest of a ridge and the enemy will find it hard both to identify targets and to hit them. The RAF group commander in the south-east, the pragmatic New Zealander Keith Park, fought the battle very much in this spirit: he used his aircraft parsimoniously and sparingly, dispatching one or two squadrons to disrupt attacks as soon as possible, rather than attempting to create large aerial formations.[4]

The battle might easily have been fought very differently: the group commander to the north, Trafford Leigh-Mallory, was a believer in assembling large forces in the air, the "Big Wing" whose most famous proponent was Douglas Bader.[5] But Park, further south, had no time to assemble such forces – to have attempted to do so would certainly have left German formations free to attack their targets unmolested. Which approach was correct is closely related to whether air combat is square law, privileging numerical advantage, or linear law, a set of duels. If Park's approach was wrong and the combat was square law, then the Luftwaffe would destroy the dispersed small forces sent to deal with German bombers and fighters. If on the other hand air combat was a set of duels, it would not matter to his forces' effectiveness if he sent up fewer planes, and he would be able to engage the Luftwaffe sooner, preventing more damage. This debate, just like the one regarding central versus distributed command in the Royal Navy, was critical to the RAF's performance.[6]

Using the data from the Battle of Britain, we can investigate *post hoc* which view of air combat – Park's or Leigh-Mallory's – is more accurate. We do so by thinking about the data's *scaling* properties. Whenever we build a model – of an animal, or a ship, or an aeroplane – we need to know how its various properties scale up or down as we increase the (length, time or other) scale. For example, a model at a length scale of 1:10 will have a volume which scales as the cube of length, and thus 1,000th of the real thing. We can do something similar for air combat by hypothesizing that each force's losses scale as some power of its own sorties multiplied by some (other) power of enemy sorties. The tactical goal, of course, is to make sure that one's own sorties tend to cause kills rather than losses. Knowing how the casualty exchange ratio scales with the numbers in action on each side will enable the commander to know whether their aircraft tend to do relatively better or worse when there are more aircraft in action.

In the Battle of Britain, the way that casualties scaled with numbers was rather different for the two sides. RAF losses were approximately proportional to Luftwaffe numbers, just as in Lanchester's model. But Luftwaffe losses, rather than scaling with RAF numbers, are instead also roughly proportional to German numbers – which is all the more surprising since, in the available data, while RAF sortie numbers are precise, the German sortie numbers are just British observations and therefore approximate. The casualty exchange ratio is therefore roughly constant, which, as noted above, is the signature of a set of duels. So a rough first approximation of attrition in the Battle of Britain is that it was a sum of duels (of the kind seen in Figure 3.2), the number of duels being roughly proportional to the number of Luftwaffe aircraft in the sky.[7]

However, there is a further subtlety, a secondary effect which is slight but robust – it reappears in air combat data over the Pacific in the Second World War through Korea and Vietnam to the Falklands. This is the asymmetry between the attacker (of ground or surface targets) and the defender.[8] The approximate results we have noted are already asymmetric – it was German, not British numbers that determined the number of duels – and this gives us a clue to what was going on. Essentially, the British took advantage of their freedom to control the extent of their engagement. Much like in guerrilla warfare, which displays the same signature scalings, Park was able to observe German numbers and respond in proportion, using small forces to disrupt attacks,

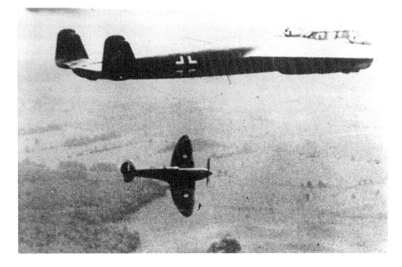

FIGURE 3.2 Duel between a German Dornier Do 17 and British Spitfire. (German Federal Archives)

ideally in swift passes without presenting British fighters as targets to Luftwaffe fighters. In the data, RAF losses scale up slightly faster than Luftwaffe numbers. Just as volume scales in proportion to the cube of length, so RAF losses scale with Luftwaffe numbers at a power that's a little greater than one. Conversely, German losses scale with their own numbers at a power slightly less than one. The two effects together result in the casualty exchange ratio not being independent of numbers, but rather scaling as roughly the square root of the number of German aircraft in the sky. The Germans did relatively better on days with large raids and large numbers, and Park's minimal approach was therefore correct. If Leigh-Mallory had been in Park's position, how much worse would the British have fared? Almost certainly much worse – but we don't have the data to be precise, precisely because Leigh-Mallory was *not* in charge.[9] However, we do have the data to think about whether the German high-level decision-making was correct.

RESCUED BY THE BOOTSTRAP

How is this to be achieved? We know that Lanchester's model is wrong for air combat: the casualty exchange ratio does not depend on force size in the correct way. Air combat is instead perhaps 80% a sum of duels and 20% asymmetric effects. But this is too loose and approximate a conclusion to be of much use in the statistical modelling of counterfactual battles. We are left without a model, with nothing to work with but the actual loss and sortie data of the campaign itself.

We are saved by a simple yet remarkably powerful statistical technique which makes a virtue of precisely this problem: the *bootstrap*. This uses no model, no insight into the process which produces the data, but simply the data themselves – it is *non-parametric*. The bootstrap is simply *resampling with replacement* from the data themselves.

It is simplest to explain this in the historical context. We have 112 days' worth of data from the Battle of Britain, with loss and sortie numbers for each side on each day (together with weather, targets *etc.*).[10] Now we accept that these data are all that we know, and all that we *can* know, about the Battle of Britain, both as actually fought and in any counterfactuals which we might propose. So, how else might the Battle of Britain have played out? We can create alternative sets of 112 days' worth of data in which each day is simply one of the actual days, chosen at random. The reason this is not merely the one actual battle is that actual days may appear more (or less) than once in the alternative set. For example, in one alternative battle, a day which favoured the Luftwaffe might appear, by chance, two or three times, while some days that favoured the British might not appear at all. The result of this alternative battle is that the Germans do a little better than in the real thing. For other sets, the opposite might be true. Now we create not just one but many such sets – 100,000, for example (which demonstrates why the bootstrap is only possible with the advent of ample computer power).

FIGURE 3.3   Kill tally of No. 303 (Kosciusko) Squadron RAF. The Poles were the most deadly pilots on either side of the battle. (Polish Institute and Sikorski Museum, London)

So far, we have been discussing aircraft (airframe) losses. The design and production in large numbers of well-armed fast monoplane fighters is of course a necessary condition for Britain's having any effective capacity to control its skies. But British fighter production was strong throughout the battle and beyond, and the critical resource was trained pilots – even though its own pilots were augmented by pilots from the Commonwealth and allies, to great effect as seen in Figure 3.3. The essential point was that for every pilot there was, and would always have been under any of our counterfactuals, an aircraft available. Pilot training too was a well-organized production line, but not one whose output could easily be ramped up in weeks or months. Henceforth, then, we deal not in aircraft losses but in pilot losses.

For our initial 100,000 re-runs of the actual battle, we plot the results in Figure 3.4. Along the base of the graph are the days of

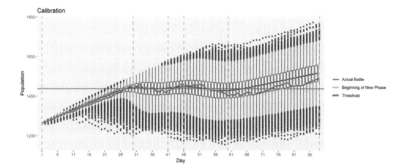

FIGURE 3.4 Comparing RAF pilot availability in the Battle as fought (red line) to counterfactual battles. These are boxplots: the boxes denote where the central 50% of the data (for a particular day in a counterfactual battle) appear. The vertical lines attached to the boxes cover 95% of the data; the circles beyond are outliers. Blue-dashed vertical lines mark new phases. The dark green horizontal line is the threshold 1,437.5, the mean of the bootstraps. As expected, the counterfactuals spread out as more days are sampled randomly but follow the general trend of the battle. (For Figures 3.4, 3.5 and 3.8–3.11, data taken from "Bootstrapping the Battle of Britain", Fagan et al. (2020). The figure was produced in R using the ggplot2 package.)

the actual Battle of Britain, with day 1 being 10 July 1940, while on the vertical axis is the number of pilots that remain. Vertical blue-dashed lines separate the different phases of the actual battle. We depict the actual battle as a red line, while "box-and-whisker" plots for each day represent the typical bootstrapped battles. The centre bar of each box is the median, or 50th percentile, below which 50% of the data lies. The edges of the box are the 25th and 75th percentiles. The whiskers form the boundaries, beyond which lie the outliers. These outliers represent some of the best, and worst, outcomes for either side, where a string of excellent or rotten luck made the difference (rather than any intrinsic difference in tactics or strategy). These variations reflect the spray of bullets, the quarter-second-too-slow reflexes and other minutiae that varied across the battle.

Of course, the actual battle constituted a British victory, in that the Germans failed to achieve air supremacy, and did not invade. The historian is free to hypothesize what made this a victory. But historians typically end up with a clear view on the margin of victory. For example, one historian might think that the actual battle was evenly balanced, with a British victory probability of 50% – that the battle was won or lost on a coin toss. A second might think that the margin was much wider, and that the British won the battle easily – with 90% probability, say. Now, any German invasion would have required the crossing of the English Channel, and a necessary precondition for this is for the Luftwaffe to believe that the RAF can no longer deploy enough pilots to contest the skies during a time when crossing the Channel is possible. But how few pilots are low enough to enable this German "victory"? Rather than argue about the intrinsic merit of any particular threshold value, we frame the question in reverse: given a historian's estimate of the probability of victory, to what threshold number of pilots does that probability correspond?

In Figure 3.5, we plot the same data as in Figure 3.4, but now with the base of the graph representing the minimum number of pilots during a possible period of invasion. The vertical axis is now the number of times this result occurred. What this figure makes clear is that the result is approximately a normal distribution, the "bell curve" in green. Using this, one can turn our historians' estimates of the probability that the British would have won into threshold pilot numbers. For example, someone who believes that the British won on the luck of a fair coin flip, 50%, would want to use approximately the mean, 1,437 pilots. On the other hand, someone who believes that the battle was won quite safely by the British, say with 90% probability, would want to use 1,367. In each case, we can simply use the bell curve to estimate a threshold. The bell curve allows us to translate any perceived victory probability into a threshold number of pilots or *vice versa*.

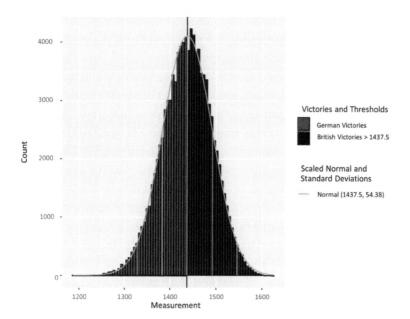

FIGURE 3.5   Comparing the results of counterfactual battles to statistics. This histogram, in which results are plotted as ranges on the x-axis and counted on the y-axis, shows how the counterfactual battles tend to cluster into a normal (bell curve) distribution. This distribution is centred at the median value for the data. If we then choose a threshold for British victory, we can convert it to a percentage by considering the proportion of battles with pilot availability better than the given threshold. Increasingly successful British victories are on the right of the figure; the best German outcomes are on the left.

A BOOTSTRAP METHOD FOR COUNTERFACTUALS

The question which our counterfactuals seek to answer is "Under what circumstances would the Germans have attempted to invade Britain in 1940?" We will not consider the possibility of a very early invasion, as soon as possible after the British forces' retreat from Dunkirk and perhaps in June, but rather an invasion as a result of German achievement of air supremacy and thus in high summer or early autumn.[11]

Our method will be to use the *weighted bootstrap*. This is the bootstrap as described above, but now with the days weighted differently, so that in the alternative, counterfactual data the probabilities of any two days are no longer necessarily equal. So rather than choose every day with probability 1/112, we might instead for example choose to double the number of days that were "like" the 16 days on which airfields were the primary targets, so that each of those 16 days is now chosen with probability 1/56. The other probabilities are then adjusted downwards so that the sum of the probabilities is still one. In this way, we can construct various counterfactual scenarios, artificially prolonging or contracting different phases of the battle or altering the Luftwaffe's surface targets.

Our mechanism, our trigger for invasion, will be the RAF pilot availability discussed above: we assume that the Germans decide to invade if the number of available RAF fighter pilots (calculated as the initial number, plus the number newly trained, minus the number lost in the bootstrapped counterfactual battle) falls below a certain threshold throughout any ten-day period beginning two weeks before a date on which the tides are suitable for invasion.[12] The principle here is that German planners need to consistently believe, over a period from a few days before an initial decision for invasion until it can no longer be cancelled, that the RAF is defeated or at least greatly weakened.

Crucially, this threshold will not be intrinsically important; rather, its job is to mediate between an opinion about the Battle of Britain as actually fought and the opinion which it is then rational to hold about a given counterfactual scenario. For our counterfactuals, we imagine three historians' views on this. The first is that the battle was entirely even, 50:50 – that the battle was won on a coin toss or, to the extent to which it was determinate, by a "narrow margin".[13] The second is that the British had an 84% chance of victory – not a round number like our earlier 90%, but not an arbitrary figure, rather corresponding to one "standard deviation" (usually written as a Greek sigma, σ) in a normal distribution, a shift of the mean by the typical amount of variation. The third view doubles this deviation, giving the British a 97.7% chance of success. (As scientists would say, for the British not to have won would then have been a "two-sigma" or "2σ" event.) As noted above, we first use these probabilities to determine associated pilot-availability thresholds for the real campaign, in a bootstrap of 100,000 battles. The thresholds turn out to be 1,437.5, 1,383 and 1,328.5. That is, for the historian who believes the battle was evenly balanced the threshold is 1,437.5, while the historians who believe in moderate and large margins of British victory believe that the British could have afforded to lose more pilots, down to thresholds of 1,383 and 1,328.5 respectively.

These thresholds are then used in the weighted bootstrap to determine three new, different probabilities of British victory for the counterfactual scenario. In a counterfactual which favours the Luftwaffe, all three figures – 50%, 84% and 97.7% – will be reduced. This way, the results make no assumptions about the reader's prejudices – you are welcome to believe that the British had, in the actual battle, a lucky victory, or alternatively that their victory was almost assured. But, given your initial beliefs, and provided you believe that the real days' outcomes are typical of those which might have occurred, the alterations in a counterfactual scenario to your beliefs about the victory probability are then determined by the bootstrap simulations. (For those who understand such things, this is something like the Bayesian approach to statistics: one begins with a "prior" belief, challenges it with evidence – here the counterfactual scenario – and arrives at a different, "posterior" belief. See the Appendix.)

It is important to note just how restrained are the counterfactuals allowed by the bootstrap, for the bootstrap uses only the days' events of the battle as it actually developed. By using only the real data, we deny all possibility that something truly out of the ordinary, something never seen in the actual days of fighting, might have occurred. There are no exceptional events, no "black swans",[14] and this is an intrinsic feature of the bootstrap – nothing (on any one day) can occur which did not happen in the actual battle.

COUNTERFACTUAL BATTLES OF BRITAIN

In 1940, the German strategic aim was to remove Britain and France from the war. The German strike through the Ardennes and drive north to the coast had split the Allied forces. This was followed by the almost-miraculous British evacuation from Dunkirk at the end of May – we won't explore the alternative of a concerted German attempt on the ground to prevent this. But at this stage, France still had to be defeated militarily, and Germany still had both to achieve this and to remove Britain from the war. Forcing the British to sue for peace was one obvious possibility.

At the time of Dunkirk, Hitler was uncommitted to the other obvious possibility, invasion. A series of meetings over the next month led to a famous order on the 16 July to prepare for invasion, codenamed Operation Sea Lion. The preparations included, among other things, extensive conversion of river barges for beach landings, as seen in Figure 3.6. The ensuing air battle has tended to overshadow other simultaneous elements of the campaign – both RAF Bomber Command and light forces of the Royal Navy conducted extensive action against the invasion fleet building up in the Channel ports and would of course have moved to centrestage had an invasion attempt occurred.[15]

For either possibility, and given the dynamic between Hitler and his air commander Goering, an air campaign was the obvious next step. The defeat of the RAF was its operational aim, but it was never clear how this was to be achieved. Attacks on ports and shipping were an obvious choice (the *Kanalkampf* which mostly took place in July). Moving on to a major assault on southern airfields in August was another. Attacking factories, most of them further inland and on the basis of imperfect intelligence, was more difficult. The most famous decision of the battle, the Luftwaffe's switch to London as its principal target on 7 September, can be seen as having a two-fold operational aim: to draw the RAF into the air there to be destroyed, and to force Britain to sue for peace by causing a collapse of morale in its

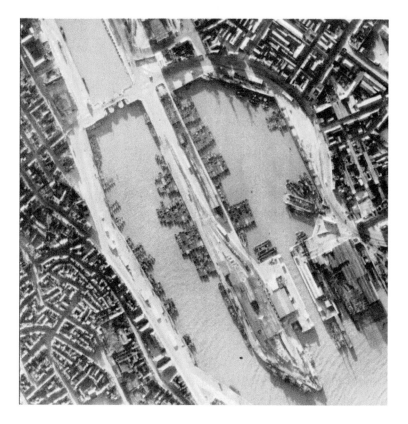

FIGURE 3.6 German barges in the port of Boulogne in preparation for invasion. (Royal Air Force, photographer unknown)

capital. This latter aim gradually came to dominate, as the battle developed into the night-bombing "Blitz", and indeed one could argue that no discrete "battle" took place at all. But the obvious operational value of simple air supremacy would have been to make an invasion of Britain in 1940 possible, and this the Germans failed to achieve.

To conduct our counterfactual analyses, we first need to say some more about the Battle, and how it developed over time. Our division of the battle into phases will be that of the official history, and the phases track the German inconstancy about which surface targets to attack. The first phase (P1: 10 July to 7 August) consisted principally of coastal attacks, attacks on shipping and armed reconnaissance. The second phase (P2: 8 August to 18 August) consisted of heavy attacks on mostly coastal targets. Later (P3: 24 August to 6 September), sustained attacks gradually concentrated on airfields: this, we know in retrospect, was when German bombing was having its greatest effect on the RAF itself, not necessarily through the effects of bombing on the ground, but rather (as we shall see in our results) through the aerial attrition that took place as the RAF sought to defend against it.[16] The final phase (P4: 7 September to 31 October) follows the switch to bombing London. We also label each day with the predominant target: A for aerodromes, C for coastal (ports and shipping), L for London and the Thames estuary, and R for mere reconnaissance. For the actual battle (A,C,L,R) = (16,47,36,13).[17]

HOW DO HITLER AND GOERING PROCEED?

We know that Hitler and Goering struggled with exactly how best to subjugate Britain. Hitler might have preferred that the British ally with the Germans, as two Germanic peoples united in global hegemony, albeit with the British maritime empire subservient to German Eurasian dominance. Perhaps a counterfactual Hitler would have seen with greater clarity the immediate strategic necessity of militarily defeating Britain, famously described by Goering as Germany's "most dangerous enemy".[18]

Goering (pictured in Figure 3.7 talking to fighter ace Adolf Galland), meanwhile, a true believer in air power, nevertheless lacked proper intelligence as to how best to use it operationally. Whether in order to drag the British to the negotiating table or to enable invasion, he had to destroy the RAF. But how was this to be done? The classic criticism of the Luftwaffe's tactics, its switch to targeting London, is a direct result of Goering's problem. What is the seat of the RAF's strength? What exactly are the RAF obliged to defend?[19]

FIGURE 3.7 Adolf Galland's fame and reputation as a high-scoring ace gave him some latitude to argue with Goering despite his relatively low rank of Major, especially concerning the latter's insistence on close escort for German bombers. (Shutterstock, Igor Golovniov)

For our first counterfactual we begin with Goering's oft-criticized mistake. We suppose that, for whatever reason, Goering chooses not to authorize attacks on London – perhaps the Luftwaffe has been receiving better combat reports from its actions over southern England. Instead, Goering continues with Phase 3 attacks. Naturally, much of the battle looks the same until this change to the actual battle occurs, at which point the RAF pilots continue on their downward trajectory, as seen in Figure 3.8. This does substantially harm the RAF's chances during the last critical period: the thresholds 1,437.5, 1,383 and 1,328.5 now mark 9.5%, 30.8% and 63.2% chances of German success. In the middle of these, the historian with a moderate belief in British victory in the actual battle – in Britain's prevention of an invasion, as actually occurred – must now have a moderate belief in a late-1940 invasion.

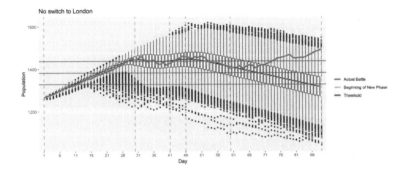

FIGURE 3.8   The counterfactual battle with no Phase 4. This figure proceeds nearly identically to Figure 3.4 but, instead of the slight downturn at the end of Phase 3 followed by a significant rise, the downturn is more impressive and continues with no reprieve for the British. The actual battle (red line) would not seem unlikely until about day 71. RAF pilot availability remains above thresholds (green horizontals, top to bottom) 1,437.5, 1,383 and 1,328.5 with probabilities of 9.5%, 30.8% and 63.2%, down from 50%, 84% and 97.7%.

Of course, Goering's decision to target London was not the only change in strategy that proved critical. Germany initially targeted Channel shipping in an attempt to isolate Britain economically, yet one could argue that this was the option least likely to achieve its goal. Goering could have instead had a much clearer conception of the operational necessity of defeating Fighter Command and a clearer notion of how to achieve it: by targeting airfields or London, both of which the RAF were obliged to defend. That is, instead of (A,C,L,R) = (16,47,36,13), Goering sets (A,C,L,R) = (30,10,62,10). What follows is a particularly brutal Battle of Britain, as seen in Figure 3.9. Even though this "counterfactual Goering" still switches to focus on London instead of the aerodromes, the damage done over land proves much more extensive than that over the Channel: the thresholds 1,437.5, 1,383 and 1,328.5 now mark 0.1%, 2.0% and 16.3% – roughly a three-sigma or 3σ shift. While London is generally considered the major mistake in Luftwaffe strategy, the *Kanalkampf* appears to have been

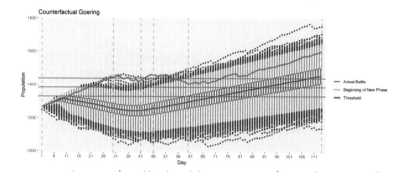

FIGURE 3.9 The counterfactual battle with better targeting of RAF Fighter Command. In contrast to Figures 3.4 or 3.8, this figure is almost immediately different with losses outpacing the initial gains that happened during the real battle. In this case, the real battle (red) would be more than just unusual; it would have been a run which briefly had the highest number of pilots for a run of days. This initial build-up was critical in the actual battle: the identified thresholds (green horizontals, top to bottom) 1,437.5, 1,383 and 1,328.5 now mark 0.1%, 2.0% and 16.3%, down from 50%, 84% and 97.7%. Hence the data suggest that targeting the Channel was a greater mistake than the Phase-4 switch to targeting London.

a greater one, especially given that it gave time for the British to continue building up their forces.

Of equal importance for the outcome of the Battle of Britain was Hitler's lack of early clarity about the strategic necessity of defeating Britain. This cost Germany dearly, as he lost the opportunity to deal the single most substantial blow that could be dealt to the RAF: an early start. If the battle had begun early, on 16 June instead of 10 July, the RAF would have had 165 fewer pilots.[20] Since the battle begins early, this also gives time for the Germans to try to take advantage of the 26 August neap tides. We assume that Goering proceeds as he had in the actual campaign, extending Phase 4 until time runs out for the Luftwaffe, as seen in Figure 3.10. Despite no other changes, this counterfactual proves to be a critical shift: the thresholds 1,437.5, 1,383 and 1,328.5 now

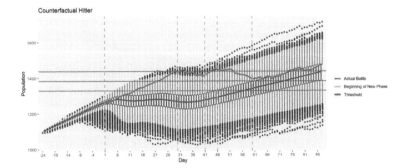

FIGURE 3.10 The counterfactual battle in which Hitler orders an early attack. Despite following nearly the same trajectory as the actual battle (red), the early assault drastically brings down the number of trained pilots that the British can field by effectively making Phase 2 the start of the battle. This results in losses much earlier than factually occurred alongside much lower numbers of pilots on day 1. The identified thresholds (green horizontals, top to bottom) 1,437.5, 1,383 and 1,328.5 now mark 0.1%, 2.4% and 16.9%, down from 50%, 84% and 97.7% and comparable to Figure 3.9.

correspond to 0.1%, 2.4% and 16.9% – again a 3σ shift, similar to the results of our counterfactual Goering. Almost all of this change results simply from the ability to use the 26 August neap tides, through a poor RAF situation around days 30–40.

A more extreme counterfactual is to combine the two above, as seen in Figure 3.11: to allow Hitler to see the strategic necessity of, and Goering to see the operational route to, the defeat of Britain. Hitler decides immediately after Dunkirk that a military victory over Britain must be achieved through invasion preceded by air victory, and Goering understands that it is RAF Fighter Command, not the British economy or the Royal Navy, that stands in the way of that decisive air victory. In this scenario, Hitler again orders that attacks by the Luftwaffe commence on 16 June and Goering obliges by taking the battle almost immediately over British ground. In this scenario, we merely scale up the

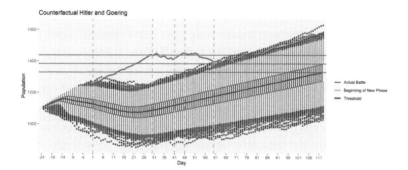

FIGURE 3.11 The counterfactual battle with an early, Fighter Command-focused attack. This figure combines the strategies of Figures 3.9 and 3.10 to brutal effect on the British. The actual battle (red) only barely touches the largest numbers of pilots obtained from the counterfactuals. This initial downturn allows the Germans a decent chance of reducing the British pilot strength to below 1,000, the only counterfactual to do so, and allows the Germans to win the aerial battle in time to take advantage of the first invasion period in every run here.

existing ratio of attacks by the additional number of days, resulting in about $(A,C,L,R) \approx (36,12,75,12)$.

Naturally, Figure 3.11 looks very similar to Figure 3.9 in its general shape. What is important here, however, is the downward shift that the overall picture experiences. Only the rarest of runs experiences the actual battle's starting number of pilots, 1,259, before September. It is no wonder then that even our lowest threshold sees no bootstraps with British victory. This is due almost entirely to the initial battering that the RAF receives: no bootstrap makes it through the August tides, had the Germans been ready to invade at this point. If they were not, then a few runs, 0.1%, make it through the lowest threshold in time for the early September tides. By the early October tides, 0.1% even manage to make it to the highest, 1,437 threshold. But invasion at some point is - unless you believe there was an overwhelming victory in the actual battle, far beyond reasonable doubt - almost certain.

We have mostly framed our results in terms of shifted probabilities, and the astute reader might note that these are very much
determined by, and relative to, the variation in the actual days'
fighting of the battle. This is one reason why scientists often prefer
to use the language of multiples of sigma, which, as noted earlier,
is the typical amount of variation. Three-sigma events do occur,
and sometimes more often than might be expected, because of
systematic deviations from true randomness. In science, this
might for example be due to the selective seeking-out and pub-
lishing of positive results. In our historical context, systematic
counterfactual changes which affect the detail of the fighting on
individual days – aircraft, fuel, ammunition, tactics, intelligence
etc. – could perhaps outweigh our two earlier, 3σ counterfactuals.
Physicists – discovering a new particle, such as the Higgs boson,
say – often operate to a 5σ standard, equivalent to a probability
of error of less than one in a million. That is why we say that our
combined counterfactual – which, very roughly (see Appendix),
adds two 3σ cases to give something in the region of a 6σ event –
takes German victory far beyond reasonable doubt.

COULD GERMANY HAVE WON
THE GERMANY OF BRITAIN?

With the probabilities that have emerged from our bootstrapped
counterfactuals, the one-word answer is clearly "yes". But this is
really "yes, if…."; and what we can gain from the quantification
achieved by the bootstrap is to place the conditions for German
victory in sharper focus – which then takes us on a journey
through an ascending hierarchy of German war aims. Let us
begin at the top.

In military theory, the overarching principle of war is always
"maintenance of the aim".[21] It is not entirely flippant to say that
Hitler's aim was world dominance, but this immediately splits in
two: the continental aim of the conquest of the east, *Lebensraum*,
a greater Germany, and a much less well-defined aim of participation in world empire. Britain had clearly got to this latter goal

first. If Britain had been the country Hitler wished – Germanic in outlook as well as in ethnicity – she could have been Germany's (subordinate) partner in this endeavour. But it is well understood that Hitler did not have a clear plan for suborning Britain into his worldview. To imagine otherwise requires that the Hitler of spring 1940, utterly clear about the need to defeat France and subdue Britain in pursuit of his greater goals, and wholly surprised by the speed and ease of his initial victory in eastern France, could have seen with a sudden new clarity of vision the urgent necessity of militarily defeating Britain simultaneously with France. A sustained German push for Dunkirk can easily be imagined, but an early Battle of Britain requires rather more commitment – for example, in moving the Luftwaffe early to Channel-coast airfields, even as the battle for France continued further west. The apparently restrained counterfactual Hitler who orders an early Battle of Britain rapidly becomes more exuberant, requiring a clarity and flexibility of mind in him which might have eluded even leaders less strait in their thinking.

At the operational level, Goering had parallel problems, with constant uncertainty about which objectives could defeat the RAF. Yet each of Goering's tactics could be said to be conceived as contributing towards the overall goal of beating England. Attacks on shipping and Channel-coast targets clearly support the invasion and benefit the Luftwaffe in being close to home bases and denying the British good observational information – and downed pilots could often be rescued, too, an activity in which the Germans were initially more effective than the British. Attacks on aircraft and engine factories – if one knew where they were and which were relevant – were clearly valuable. With its combat experience and the Me109, which was especially effective when operating from height, as its principal fighter, for the Luftwaffe to attack any target which the RAF must come up to defend, and thereby defeat Fighter Command in the air, could plausibly appear a winning operational strategy – and indeed was the rationale behind the switch to attacking London. Fighting over enemy territory, it was very hard for the Germans to know what worked, and we might argue that on both sides it was *information* (or the lack of it) which was the decisive factor in the battle.[22] In contrast, the British system – largely created by one man, Hugh Dowding – of digesting information from radar, observation and radio reports and turning it into precise, actionable information for fighter squadrons was what enabled the careful use and husbanding of resources. On the German side, there was next to no information, and this led to dissipation of effort. To posit our counterfactual, Goering requires the Luftwaffe to have planned for and won the information battle, and to have had a clear operational conception of how to use an air force to achieve strategic goals when not operating in close support of ground forces. This is a strong counterfactual in the history of military ideas: neither the classic theory of land war nor the newly arrived theories of Douhet on strategic bombing provided a framework for thinking about how to defeat an air force. Unless we allow the Germans some lucky guesswork, our counterfactual Goering requires him to win a war of information and ideas for which the Germans simply had not prepared. But it is an important point that their contemporary opponents had done so thoroughly, that the system would not have existed but for one man and that they used their system to win.[23]

Thus, in the cases of both Hitler and Goering, what began as a restrained counterfactual was pulled by the data towards something stronger, amounting to a single person acting and deciding differently in ways which, while perfectly feasible, would have been out on a limb of the tree of possibilities. Is positing a counterfactual mindset in an individual, or their replacement by a different individual, visionary beyond the norms of their times, restrained or exuberant? Such individuals certainly exist and attain positions of power, and it is surely a hindsight error to claim that this is exuberance merely on the basis that the decision-making individuals were who they were. But this is not a question to which we can give a quantified answer, and it perfectly exemplifies the tensions in "great man" history.[24]

We can, however, say now with some quantified justification that, for most historians and certainly for those who believe in the "narrow margin", it was materially possible for Britain to have lost the battle, and therefore to have been invaded – although whether an invasion could have succeeded is another question entirely. Moreover, bootstrap methods enable us to quantify comparisons of contrasting views on differing decisions, providing a jumping-off point for qualitative analysis rather than reducing the debate to mere clashes of opinion. This is what mathematics does, of course – its truth is found in the argument connecting assumptions to conclusions, not in the conclusions themselves.

We have begun to see in this and the previous chapter that crunching the numbers in battles of attrition can throw into sharper focus a range of deeper questions stretching back years. Creating and manning a military force requires funding, national will and a clear conception of the tactical necessities of its use. When war arrives, simple attrition brings a new perspective not only on tactical and operational decisions but also on the deeper questions of the interplay between personalities and strategic aims.

In the next chapter, we take this to the next level, to a war in which, for the first time, computer power was allied to the gathering of a huge quantity of data and, in contrast to the historical perspective of the previous chapters, contemporary decision-makers were conscious of this data and had the will to use it. Yet this was also a war which would finally depend more on two societies and their politics than on data or the use it was put to: Vietnam.

NOTES

1. Lanchester, *Aircraft in Warfare*.
2. Efron and Tibshirani, *Introduction to Bootstrapping*.
3. *The Narrow Margin* is the title of the classic study by Wood and Dempster. For an exploration of the myth and how it inverts aspects of the truth, see Bungay, *Most Dangerous Enemy*, which is an excellent general introduction, used by the RAF in its officer cadet

course. Townsend's *Duel of Eagles* is particularly thoughtful in tactical matters.

4. An excellent biography is Orange, *Park*.

5. Sarkar, *Bader's Duxford Fighters*.

6. For a narrative history of the debate, see MacKay and Price, *Safety in Numbers*.

7. Johnson and MacKay, *Lanchester Models and the Battle of Britain*.

8. Horwood, MacKay and Price, *Concentration and Asymmetry in Air Combat*.

9. It is worth noting that Leigh-Mallory's fighter sweeps over France in 1941 was unsuccessful in precisely the way we would expect. However, in these, the roles of attacker and defender (of the territory below) were of course reversed.

10. James, *Battle of Britain*. We do not include 25 September (an attack on Filton) and 16 October (German air sweeps), which are inappropriate for inclusion in the bootstrap. Note the five-day lull between (P2) and (P3); we treat this lull as a separate phase (P0) in our scenarios.

11. If the Germans had invaded England, the decisive question would have been whether the Royal Navy could have denied the English Channel to the German invasion and re-supply fleets. This perspective is given in Grinnell-Milne, *Silent Victory*. The invasion is wargamed in Cox, *Operation Sea Lion*. A classic counterfactual early invasion is Forester, *If Hitler had invaded England*; see also Macksey, *Invasion*. For German planning and conditions for invasion, see anon., *German Plans* and Schenk, *Invasion*.

12. To be precise, we take the British pilot strength to be $1259 + 9.35 \times T - L$, where 1259 is the number of trained pilots available on 6 July 1940, T is the number of days beyond that date, 9.35 is the number of new pilots per day (11) reduced by a factor of 0.85 to account for Dowding's belief in the lesser value of a novice and L is the number of pilots lost in the bootstrap simulation. The trigger for invasion requires that this be below threshold throughout the period from $T - 15$ to $T - 5$ for all T between $Q - 3$ and $Q + 3$, where Q is the quarter-moon, thus giving the neap tides suitable both for beaching just before dawn and for aiding the movement down-channel of the invasion fleet.

13. This is famously part of British mythology about the battle; the classic history is Wood and Dempster, *The Narrow Margin*, which formed the basis of the 1969 film "Battle of Britain".

14. Taleb, *Black Swan*.

15. Grinnell-Milne, *Silent Victory*.

16. Whilst one could legitimately argue that under more intense assaults bases such as Biggin Hill might be destroyed, this is irrelevant for the bootstrap as long as the skies are still contested – almost certainly true as the RAF would attempt to maintain even temporary forward bases as close as possible to any landing site.

17. Data sources: airframe losses and pilot casualties are from Ramsey, *Battle of Britain*, and Wynn, *Men of the Battle of Britain*; targets are from Hugh Dowding, *Enemy Air Offensive Against Great Britain, 1941–1947*, held as AIR 2/7771, UK National Archives, Kew; sortie numbers are from James, *The Battle of Britain*. German sortie numbers are not generally available before 1 August.

18. See, for example, Stephen Bungay's eponymous book.

19. This criticism is made by airpower theorists such as Warden, *The Air Campaign*.

20. The attack could not have been brought forward by much more than a month, given the pattern of meetings between Hitler and his commanders. Further, the conquest of France needed to be sufficiently far advanced to leave Channel-coast airfields in German hands and then made operational.

21. Such principles were first posited in Fuller, "The Principles of War".

22. The asymmetry between attacker and defender (of surface targets) noted in Horwood, MacKay and Price, *Concentration and Asymmetry in Air Combat*, is in part an information asymmetry: the defender usually knows where the attackers are and where they are going.

23. This is true to such an extent that it is possible to write a paper with the title "An information system won the war" (Holwell & Checkland).

24. In the cases of both Hitler and Goering, what we are doing here is to pose the first of historians' three natural challenges to "theoretically subversive counterfactuals", to "challenge the mutability of the antecedent". Tetlock and Lebow, *Poking Counterfactual Holes*.

CHAPTER 4

Could the United States Have Prevailed in Vietnam?

FIGURE 4.1   American soldiers of the 101st Airborne Division during Operation Van Buren, South Vietnam, 1966. American ground forces assumed increasing responsibility for the allied war effort after 1965. (U.S. National Archives and Records Administration, Signal Corps photographer)

DOI: 10.1201/9780429488405-4

HISTORICAL CONTEXT

We now move on another 25 years, to South-East Asia in the 1960s. The French withdrawal from Indochina had left Vietnam split between a communist North and a western-backed South, which the United States was trying to prevent the North from annexing. The war was a mix of conventional military campaigns and insurgency, played out against a counterpoint of political commitment or its absence. But how should the United States and its allies balance war-fighting with peace-making, attrition with pacification? And how could the United States' decision-makers best collect and use data, and use newly available electronic computers, to answer this question?

After a decade of war, defeat in Vietnam came as a profound psychological blow to Americans. It initiated an open-ended *post mortem* about what had gone wrong in South-East Asia that persists to this day. Much of this debate has involved the ventilation of competing views as to which alternative strategy would have produced a conclusion more favourable to the United States and its South Vietnamese ally. Thus the war has lent itself, perhaps to an unusual degree, to the creation of counterfactual solutions to the conundrum of US defeat. Clearly, the strategy adopted by the United States in Vietnam did not produce the results desired by Washington, and there are reasons for thinking that it was flawed from the outset. It is against this historical strategy that all counterfactual alternatives must be measured. Most of them can be dismissed as unrealistic propositions, but there are reasons for cautiously advocating an alternative strategy based on an increased effort to ensure the security and pacification of the South Vietnamese population in the countryside. We discuss here how this strategy might have offered the prospect of an improved outcome for the allies and would constitute a restrained and rational counterfactual scenario.

Vietnam proved a puzzling and frustrating experience for the United States. Unlike most of its major military campaigns, Vietnam was a "war without fronts".[1] Progress could not be measured in terms of advances and retreats, of lines on a map. Instead, the United States tried to establish metrics by which it could measure its progress and on which to base operational and tactical decisions. Consequently, United States Secretary of Defense Robert S. McNamara exhorted those agencies concerned with the war to measure everything about it that could be measured and this they most certainly did.[2] If lack of information was an ever-present flaw in the German attempt to pacify the British Isles, the same could not be said of the American attempt to suppress the Communist insurgency in South Vietnam. Never in the history of American arms had so much data been collected about one of the nation's wars and never before had the application of data assumed such centrality in strategy formulation and justification. The accuracy and usefulness of this statistical windfall has been the subject of considerable debate.[3] However, conditional on the consistency of the data with the historical record, we argue that it can be used to propose and assess the viability of alternative "counterfactual" allied strategies based on pacification and more broadly establish some of the ways in which US quantification of its war effort affected its actions and the outcome of the war.

THE SECOND INDOCHINA WAR

The United States became closely involved in Indochina towards the end of the Second World War, when it began to provide support for nationalist Viet Minh guerrillas in exchange for their assistance against the occupying Japanese. However, at the end of the war the United States terminated its relationship with the Viet Minh – who were dominated by the Indochinese Communist Party – and actively assisted France to return to its former colonies in the region. As the Cold War deepened, the United States became increasingly committed to the support of France in its colonial war with the Viet Minh that erupted in 1946, but despite this assistance France was finally defeated. The ensuing Geneva Accords of 1954 split Vietnam into northern, communist and southern, non-communist "regroupment" zones. Although the Accords also called for national elections to reunify the country in 1956, no such elections were held, and the two zones crystallized into *de facto* independent states. The United States now became the principal sponsor of the southern, non-communist, Republic of Vietnam as a bulwark against communist expansion in the region.

At the end of the 1950s, a communist insurgency began in South Vietnam. This was a continuation of unfinished business from the First Indochina War, with the insurgents of the National Liberation Front (NLF), heirs of the Viet Minh, seeking to unify all of Vietnam under communist rule. The United States responded with the application of various counter-insurgency techniques to South Vietnam, supported by American advisors and specialist military units, and hardware such as helicopters and armoured vehicles. Although the South Vietnamese and their American allies seemed to enjoy some periodic success, it was clear by 1964 that the war was not going well, with increasing infiltration of Communist North Vietnamese logistic support and personnel to the South and increasing amounts of South Vietnamese territory falling under insurgent control. In the face of deteriorating conditions in South Vietnam, President Lyndon Johnson (pictured in Figure 4.2 with COMUSMACV General William Westmoreland) was obliged to consider escalating the US role in the war or cutting its losses and withdrawing support for the southern regime. Following a brief period of internal debate, the Johnson administration took the decisions to commence the systematic bombing of North Vietnam and deploy American ground forces in South Vietnam. The American Military Assistance Command Vietnam (MACV), which had been tasked with providing advice and support to the South Vietnamese government, was now transmuted into a full-blown

FIGURE 4.2   Commander, US Military Assistance Command, Vietnam (COMUSMACV) General William C. Westmoreland with President Lyndon B. Johnson at the White House. Westmoreland decided that the best use of US military power was against the more conventional part of the Communist forces, the NLF main forces and, increasingly as the war drew on, the North Vietnamese People's Army of Vietnam (PAVN). (U.S. National Archives and Records Administration)

military headquarters, and the United States assumed principal responsibility for the prosecution of the war.

The Vietnam War was an extraordinarily complex affair. Often described as a guerrilla insurgency, the war bore characteristics of both a low-intensity irregular conflict and a limited conventional war. This confronted the United States with a conundrum: which "war" should take precedence? Or should both wars be given equal attention?

In early 1961, the NLF established the People's Liberation Armed Forces (PLAF) as a single command for its insurgent forces in South Vietnam. The PLAF combat forces were organized according to the triad structure employed by the Communists since the First

Indochina War. Self-defence forces were organized at the local level and were civilians during the day and activists at night conducting various logistics and terrorist activities; guerrilla forces were organized at the regional level and conducted full-time guerrilla activity; and the main forces were regular troops drawn from the best elements of the guerrillas and formed into battalions.

MACV's new commander (COMUSMACV) General William Westmoreland, who assumed the post from General Paul D. Harkins on 1 June 1964, needed to establish an allied strategy that would accommodate the influx of American troops into South Vietnam and bring victory to the allies. Westmoreland decided that the best use of US Military power would be against the more conventional part of the Communist forces, the NLF main forces and, increasingly as the war drew on, the North Vietnamese People's Army of Vietnam (PAVN) units deep in the jungles and highlands near the Cambodian and Laotian borders.

Westmoreland was convinced that the presence of these forces in South Vietnam posed an existential threat to the survival of the country in the shape of a potential conventional offensive that could penetrate the main population centres in the coastal regions and around Saigon. His strategy would eliminate this pressing danger by bringing the United States' overwhelming firepower to bear where it would do the least damage to the South Vietnamese population. It would also enable the United States to bleed the Communist forces, highlighting the costs to North Vietnam of supporting the war in the South. Given the nature of the Communist main forces Westmoreland hoped to destroy, this strategy would become popularly known as the "big unit war". Meanwhile, the South Vietnamese Army of the Republic of Vietnam (ARVN) would look after security and pacification operations against the Communists in the more populated areas.[4]

While there is no doubt that Westmoreland concentrated on the conventional threat to South Vietnam, Secretary McNamara encouraged him to put more American effort into pacification, though with only limited success.[5] "Pacification" potentially involved a host of different strategies and approaches about which there was little firm agreement among the allies. These ranged from military security tasks to political and economic initiatives – the so-called "hearts and minds" policies designed to encourage the support of the rural population for the Republic of Vietnam government in Saigon. McNamara's concern for this aspect of the war was reflected in an effort to unite the multiplicity of different American military and civilian programmes that might fall under the umbrella of pacification. President Johnson, therefore, on 9 May 1967, authorized the combining of all these programmes under a new agency entitled Civil Operations and Revolutionary Development Support (CORDS) to be headed up by a civilian but which would be under the authority of MACV.

Westmoreland's strategy was designed not only to provide security behind which the South Vietnamese could go about the more mundane tasks of fighting a counterinsurgency war but also to defeat North Vietnam by generating Communist casualties at a faster rate than they could be replaced with fresh forces from North Vietnam or recruits from the south. Thus Westmoreland's strategy was "attritional". It did not rely on the obviously measurable seizure of geographical objectives like towns and cities for, insomuch as such geographical features might be said to be potential objectives, they were mostly already in the hands of the South Vietnamese government. Rather, Westmoreland's intention was to destroy the main force units, which essentially involved killing their component soldiers. A critical "crossover point" would be reached when the rising tide of Communist casualties exceeded the declining number of infiltrators, after which Communist strength would begin to decline in South Vietnam. Somewhere after this point, Westmoreland believed, Hanoi would realize that the game was not worth the candle and would take the entirely rational step of withdrawing from the lists, cutting off vital support to the Southern insurgency. With North Vietnam out of the war, the ARVN would find it that much easier to deal with the declining pool of unsupported insurgents in South Vietnam and

FIGURE 4.3   Dead Viet Cong soldiers at Tan Son Nhut Airbase, Saigon, during the Tet Offensive in 1968. General Westmoreland sought a "crossover point" after which the Communists would not be able to replace their battlefield losses. (Vietnam Center & Sam Johnson Vietnam Archive)

the war could be ended by negotiated settlement or the outright military defeat of the insurgency.[6] In order to judge the success of this attritional strategy, it would be necessary to measure the losses of both sides and it is from this requirement that the notorious "body count" emerged (see Figure 4.3). As we will discuss, this is an extremely hard metric to quantify, leading to a potential problem of recording only garbage data.

Such was the essence of the allied strategy from 1965 until early 1968, when the Communists mounted a major offensive during the Tet lunar new year holiday. The Tet Offensive was a deeply shocking event for the United States. The allies' response to the offensive was broadly successful; except for Hue city, in northern

Thua Thien Province, all the Communist objectives were back in allied hands within a few days and Hue itself within a matter of weeks. The cost to the Communist forces was enormous and the NLF never fully recovered, obliging the PAVN to take over increased responsibility for the war in the south. However, the US Government had led the American public to believe that the war was being won, a point reinforced by Westmoreland himself in two visits to the United States in 1967, made specifically for the purpose of stiffening American resolve as the US commitment became more open-ended. The sight of NLF sappers in the very grounds of the US Embassy in Saigon was entirely at variance with Westmoreland's speech at the National Press Club on 21 November 1967 in which he said the United States had reached "an important point [in Vietnam] where the end begins to come into view".[7]

The Tet Offensive represented a watershed in American Vietnam policy. It prompted Westmoreland, at Army Chief of Staff Harold S. Wheeler's urging, to seek a further 206,000 American reinforcements in addition to the half-million American troops already deployed in South Vietnam, in order to go on the counteroffensive. President Johnson expected to comply with Westmoreland's request, but before doing so he ordered his new Secretary of Defence Clark Clifford – Robert S. McNamara having left office at the end of January 1968 – to assess the troop request. This investigation convinced the previously hawkish Clifford that Vietnam was a losing proposition. Clifford's advice to Johnson was, therefore, to find a way to withdraw US forces, presumably on the back of a negotiated settlement.[8] Westmoreland was recalled to the United States in June 1968 and the reduction of American forces in South Vietnam began. Before Tet, Americans thought that they could win the war in Vietnam, and that they were doing so. After Tet, although the war was to go on for another seven years, the United States was essentially managing defeat by withdrawing in such a way as to preserve what President Richard Nixon would call "peace with honour" in the Paris Peace Agreement of 1973, which effectively ended US direct involvement in the war. The collapse of South Vietnam in 1975 in the face of a new Communist offensive made clear the reality of the United States' defeat in Southeast Asia.

FACTUAL VIETNAMS

In the years since the Vietnam War, the bulk of American historical writing on the conflict has contributed to a consensus that the war was both immoral and unwinnable on the part of the allies.[9] According to this "orthodox", or liberal, school of thought, there is no point in considering how the war might have been more effectively pursued – that is, no point in trying to consider counterfactual strategies – because, in this view, the objective circumstances of the war render such an effort sterile. The other main school of Vietnam War historiography, the "revisionist" school, is built on counterfactual arguments explaining how the United States could have won the war had it adopted alternative strategies to those pursued by its generals and politicians. It is, essentially, across the fault-lines of these different strategies that much of the historiography of the war is based.

Revisionist alternative strategies come in two main flavours, involving an emphasis on either conventional military operations or increased pacification. The former includes a more intense bombing of North Vietnam, an amphibious invasion of North Vietnam and an incursion into Laos to cut the Ho Chi Minh Trail and isolate the southern insurgency from North Vietnam. The validity of these counterfactuals has, however, rarely been subjected to rigorous critical analysis. Until relatively recently, the revisionist school contained few professional historians; its members tended to be serving or retired military officers.[10]

For their part, professional historians have been reluctant to engage with what "might have been" in Vietnam; not only were alternative Vietnam strategies fundamentally ahistorical, but it is also especially difficult to analyse them in an objective manner. Professional historians have, consistent with the nature of their trade, preferred instead to concentrate on interpreting the historical record, or as much of the historical record as they can uncover or restore. Here we seek to redress this imbalance between stated counterfactual "roads not taken" and their critical analysis, by addressing ourselves to some of the wealth of data arising from American efforts to assess the progress of their campaign in Vietnam. The results may establish the extent to which the United States was pursuing a "strategy for defeat" and how plausible variations of that strategy might have changed the outcomes. Naturally, we need to appreciate the reality of the Vietnam War strategy, in order to illuminate any alternatives. This means that we must first understand the detail underlying the attritional big unit war pursued by the allies before we begin to explore the counterfactuals.

As we have seen, central to Westmoreland's attritional strategy was the body count, and this proved a problematic metric from the outset. There were serious issues with collecting accurate data for enemy killed in action. It required extensive policing of the battlefield after contacts had occurred, for the Communists tended to carry their casualties away after an action and the number of weapons recovered did not necessarily match the body count. Furthermore, there were many incidents, following artillery and air strikes for example, where body counts could only be estimated and there was, of course, also a strong incentive to game the system by overestimating, since a high body count was regarded as a measure of individual military personnel's success.

Furthermore, while the attritional strategy certainly killed many Communists, the body count statistics did not necessarily reflect any real improvement in the fortunes of the allied war effort. Body count neither corresponded to the level of control exercised by the South Vietnamese government on the countryside nor had any direct impact on the infiltration of Communist troops and supplies into the South. Attrition seemed not to affect the North Vietnamese diplomatically either. US planners assumed that the North Vietnamese were "rational actors" who, if shown that the United States was sufficiently committed to the destruction of enemy main forces in South Vietnam and the gradually escalating bombing of their homeland, would be convinced that the war could not be won in the South and the costs to North Vietnamese development were prohibitive.[11]

Perhaps the most public affirmation of this view of the North Vietnamese as rational actors was President Johnson's proposal to finance a Mekong River equivalent of the American Tennessee Valley Authority hydroelectric scheme as an irresistible inducement for North Vietnam to end its support for the southern insurgency and cooperate with the United States in the development of the region.[12] In fact, however, North Vietnam's leaders' dedication to the cause of Vietnamese nationalism mystifyingly exceeded the rational economic concerns assumed of them by the Johnson administration. Clearly, the Johnson administration had misinterpreted the war aims of their North Vietnamese opponents and so had placed themselves in what was effectively a counterfactual mindset where they imputed attitudes (American "rationality") to their opponents that they might not have shared themselves had they been placed in similar circumstances, a theme we will return to in the next chapter.

Planning based on these erroneous assumptions about Communist motivations set the allies up for failure. The attritional strategy demanded an intensive approach to the war with large numbers of American expeditionary troops and the application of massive firepower which alienated the local population and rendered the war less politically sustainable for the United States than a more limited intervention might have been.

Limited success on the battlefield and its own domestic political situation obliged the United States to abandon the attritional strategy before they considered it had paid off. There are, however, good reasons for thinking that the attritional strategy would never have paid off, involving as it did many faulty assumptions about North Vietnam's ability and desire to sustain the Communist campaign in South Vietnam.

In its raw form, using the body count as a criterion for victory ignores the fact that most military strategies assume casualties on the part of friendly forces: what if the enemy completes its objectives before it runs out of troops?[13] This raises the fundamental issue of the size of the enemy force, aside from the attrition rate. Even if the enemy is haemorrhaging troops, if it has a lot to begin with then these may take a very long time to run out. Here, the endurance of the opponent becomes a crucial factor. Would the United States be able to sustain its effort in Vietnam until the Communists ran out of troops? To judge this, it was necessary for MACV to establish just how many troops the Communists had, and this proved difficult.

One problem was the unconventional nature of the Communist forces: it was not necessarily entirely clear which forces should be counted. US analysts divided them into four categories, which conformed largely to the tripartite structure previously defined plus a politico-military infrastructure in the towns and villages. We have already discussed the main forces, who could clearly be understood to be soldiers. The administrative service forces provided vital command and control functions. The irregulars included the guerrilla Viet Cong plus village self-defence forces and clandestine forces which conducted agitprop and intelligence operations in areas nominally controlled by the South Vietnamese government. Should all these forces be estimated? MACV thought so until October 1967.[14] Also, on whose estimates should United States' strategy be based? Both MACV and the CIA produced estimates of Communist strength by different methodologies which yielded markedly different results.

In the Spring of 1967, Westmoreland and his staff at MACV became increasingly convinced that the crossover point had been reached or was close at hand, suggesting that, according to the body count criterion, the United States was winning the war.

Westmoreland told President Johnson so on a trip to Washington in April 1967, and in August 1967, Colonel Daniel Graham, an intelligence analyst with MACV and a former CIA analyst, produced a memo in which he argued that the crossover point had been reached in June of that year.[15]

In contrast to this MACV optimism, in June 1967 the CIA complained that MACV was seriously underestimating the strength of Communist irregular forces in South Vietnam and it was as a result of this debate that MACV took the decision, in October 1967, to drop some of the categories from their calculations for a Special National Intelligence Estimate.[16] Acceptance of the much higher CIA estimates would involve an admission by MACV that things were not going so well, that perhaps the crossover point had not been reached, or if it had Communist forces would take many years to grind down at current rates, or perhaps such figures indicated that the United States was losing the war. MACV pressed for a compromise, which was eventually, reluctantly, accepted by the CIA. This revised estimate dropped the self-defence forces and the politico-military infrastructure but included higher CIA estimates for administrative service and guerrilla forces with the result that the total estimate of Communist forces had fallen from 294,240 to 235,852, whereas if the deleted categories had been retained, Communist forces would have risen to 395,025, or even to 482,452 in the CIA worst-case analysis.[17]

The new estimates represented instant progress, but it is not quite true that Westmoreland deliberately deceived the Johnson administration in order to maintain his claim that the war was being won, an accusation directed at Westmoreland in a television documentary of 1982, *The Uncounted Enemy: A Vietnam Deception*, over which Westmoreland sued CBS.[18] The easy availability of the conflicting order of battle estimates and the open nature of the debate between MACV and the CIA indicate, as McNamara insisted to CBS, that there was no conspiracy to deceive the US Government.[19] Westmoreland's rationalization for excluding the order of battle categories – that they were not combat forces as such and were difficult to count – was true up to a point. The fact is, though, that Westmoreland had deemed them worth counting in the past, until the higher CIA figures threatened MACV with a "public relations disaster".[20] COMUSMACV was genuinely sceptical of the CIA figures, and in truth they turned out to be highly inaccurate, but he was also adjusting the figures in an arbitrary manner to suit the established narrative that the US effort in South Vietnam was making progress, regardless of the statistical evidence on which that progress was supposed to be based.

Despite his close association with the attritional strategy, McNamara himself became increasingly disenchanted with it. In November 1966, using CIA and Department of Defence assessments, McNamara opined that "… the data suggest that we have no prospects of attriting the enemy force at a rate equal to or greater than his capability to infiltrate and recruit, and this will be true at either the 470,000 personnel level or 570,000 … it does not appear that we have the favourable leverage required to achieve decisive attrition by introducing more forces".[21] In McNamara's view, the United States could never reach the crossover point and McNamara advised that the President should consider capping the size of the US effort in-country and seek a negotiated settlement. McNamara continued to argue this point in government – but not in public – until he left office just before the Tet Offensive of 1968.[22]

According to a report by the Southeast Asia Office, Assistant Secretary of Defence (Systems Analysis) on North and South Vietnamese manpower, published in August 1968, taking the worst Communist loss rates of the war in the first half of 1968, including the Tet Offensive, if these were repeated year-on-year, a rough calculation suggested that the Communists might be able to sustain their war effort for another 12 years or so.[23] But the United States could not hope to sustain its war effort at current rates for this length of time, and in any case this level of Communist casualties was never repeated.

COUNTERFACTUAL VIETNAMS

Could a counterfactual increase in US troop strength, and thereby attrition of the Communist forces, have won the war if it had been recommended by Secretary of Defence Clifford after the Tet Offensive in 1968? While allied forces always outnumbered Communist forces during the war, the ratio was never significant enough to grind the Communists down and the Communists maintained the tactical initiative. As *National Security Action Memorandum 1—The Situation in Vietnam* summarized at the end of 1968, "there is general agreement with the JCS statement, 'The enemy, by the type of action he adopts, has the predominant share in determining enemy attrition rates.' ... CIA notes that less than one percent of nearly two million allied small unit operations conducted in the last two years resulted in contact with the enemy and, when ARVN is surveyed, the percentage drops to one tenth of one percent".[24] In practice, this meant that the Communists could decide whether to take casualties or not, just as the Royal Air Force chose when and how to engage the Luftwaffe in Chapter 3.

Indeed, even in 1962 modellers were aware of this problem. Seymour Deitchman, who would later work in the Department of Defense, had demonstrated it by modifying Lanchester's model to account for guerrilla combatants.[25] As mentioned previously, Lanchester's model for modern combat had two cases: aimed fire, which has losses in proportion to just the number of opponents and leads to a square law, and unaimed fire, which has losses in proportion to both combatants and leads to a linear law. Now consider a search-and-destroy operation, where small groups of allied troops are searching for hidden groups of insurgents. The resulting model mixes aimed fire from the insurgents (to whom the allied troops are highly visible) with unaimed fire from the allies (who first must find the insurgents). The upshot is a mixed law, in which the insurgents' fighting power is proportional to their numbers (a linear law) but the allies benefit from the square of their numbers (a square law). On the surface, the allies could benefit from a troop surge.[26] More subtly, however, the insurgents benefit from a more easily controlled density effect, allowing them to "melt away" to control their losses. Greater forces would not bring greater initiative. Even if a troop surge did allow more forces to engage, other manpower requirements, *e.g.* logistics, meant that a claimed manpower advantage that ranged up to 6–1 was in practice more like parity.[27]

The foregoing combined with the aftermath of the Tet Offensive suggests that the deployment of additional US Army units in the big unit war would have had negligible impact on the Communist main forces. We can, therefore, suggest that attrition was never a viable strategy. In practice it was largely abandoned in the late summer of 1969, as the principal American strategy shifted to Vietnamization, which represented an American admission of defeat disguised as the search for "peace with honour". Increasing the number of American troops involved in attrition would have been unlikely to have made much difference except for probable increased American casualties, intensified anti-war protest, and a weakened American ability to keep troops in Southeast Asia. Assistant Secretary of Defence Alain Enthoven, writing about Westmoreland's post-Tet request for 206,000 more troops, made precisely this point on 20 March 1968: "the notion that we can 'win' this war by driving the VC/NVA from the country or by inflicting an unacceptable rate of casualties on them is false. Moreover, a forty percent increase in friendly forces cannot be counted upon to produce a forty percent increase in enemy casualties if the enemy doesn't want it to happen".[28]

The inherent flaws in the case for attrition also raise intriguing questions about the Secretaries of Defence. One thing that is perhaps surprising about both McNamara and Clifford is their continued dedication to a metric that both came to see as a faulty indicator of allied success. The former developed deep doubts

FIGURE 4.4 US Secretary of Defence Robert S. McNamara in February 1968. Despite his implementation of the Hamlet Evaluation System (HES), McNamara, like his replacement as Secretary of Defence, Clark Clifford, seems to have been almost as obsessed with attrition as General Westmoreland. (U.S. National Archives and Records Administration)

about the attritional strategy, yet in his memoirs his assessment of the war is expressed almost entirely in terms of attrition; he did not think attrition would work but seems to have been unable to conceive of the war in any other way. Similarly, when asked to assess the prospects for US victory after Tet, Clifford's review seems to have been almost exclusively concerned with one metric – body count – and despite McNamara's exhortations to measure everything about the war that could be measured, no other metrics – such as those generated by the Hamlet Evaluation System (HES) relating to pacification – seem to have featured in the deliberations of the reviewers.[29] Indeed, McNamara and Clifford seemed almost as obsessed with attrition as Westmoreland (see Figure 4.4). The only difference is that they both came to appreciate that an attritional strategy would not work in Vietnam.

While Westmoreland pursued an attritional strategy within the confines of South Vietnam until 1968, the geographical limitations of this strategy were forced upon Westmoreland by his political masters. In March 1966, Army Chief of Staff General Harold K. Johnson produced a report entitled *A Programme for the Pacification and Long-Term Development of South Vietnam* (PROVN). This report has been regarded as a rejection of Westmoreland's strategy of attrition in favour of a strategy based on the pacification of the Vietnamese countryside, but Westmoreland found himself in agreement with aspects of the report, including Johnson's assertion that "the bulk of US-FWMAF (Free World Military Assistance Forces) and designated ARVN units must be directed against base areas and against lines of communication in SVN, Laos and Cambodia …".[30] In the *Report on the War in Vietnam* Westmoreland wrote with Admiral US Grant Sharpe, on the occasion of his recall, Westmoreland made it clear that his favoured strategy was to operate directly against not only Communist bases in South Vietnam but also the Communist sanctuaries in Cambodia and Laos, and against North Vietnam directly.[31] Indeed, in the immediate post-Tet environment, Westmoreland thought that he might have the option of carrying the allied war effort into the Communist sanctuaries, and this lay behind the request for an additional 206,000 troops. However, the Johnson administration had always been reluctant to undertake the necessary mobilization of the US reserves, lest it interfere with the President's Great Society domestic reform programme. Johnson's advisors were also concerned that it might bring China into the war, a development which the American people had shown little enthusiasm for during the Korean War. In the event, of course, the request for 206,000 troops did not survive the Clifford review, and in any case an invasion of the north would likely present the United States with a bigger problem, even ignoring the difficulties with justifying the counterfactual itself. Even if such an operation went well and the United States managed to occupy all or part of North Vietnam, they would have increased the amount of population and land the allies would need to pacify and defend. Meanwhile, the Communists could continue to use bases outside Vietnam, including China, and the likelihood of Soviet intervention would also have been significantly increased.[32]

Plans for an incursion into Laos were famously promoted after the war as a counterfactual alternative to attrition by the revisionist US Army officer Colonel Harry G. Summers in his 1982 book *On Strategy*. Such a proposal had been rejected by Army Chief of Staff Harold Johnson in 1965 because of the extensive support requirements for operations in undeveloped Laos. Andrew Krepinevich, himself a revisionist of the counterinsurgency variety, has argued that a Laotian incursion would likely have not worked because until the 1968 Tet Offensive, the southern insurgents were mostly dependent on southern sources for their logistic requirements, not North Vietnam, and in any case they would likely have shifted the Ho Chi Minh trail eastwards through Thailand, though this would have required an enormously extended line of communication. Furthermore, says Krepinevich, deploying US troops exclusively in Laos, rather than in South Vietnam, as Summers proposed, would have denied their support for the ARVN in South Vietnam, accelerating their decline, which was the very reason for US intervention in the first place.[33]

These intensified conventional strategies would, in our opinion, constitute counterfactuals of such exuberance that they can be dismissed as non-viable. There are, however, reasons for thinking that the early employment of an alternative strategy involving an increased emphasis on pacification would constitute a more restrained and rational counterfactual scenario.

During the war, and subsequently, critics such as US Ambassador to South Vietnam Henry Cabot Lodge Jr., Army Chief of Staff General Harold K. Johnson, Commandant of the Marine Corps General Wallace Green, the commander of the III Marine Amphibious Force in South Vietnam, General Lewis Walt and many of Westmoreland's subordinates argued that Westmoreland was pursuing the wrong war. The "real war", they argued, was the pacification struggle in South Vietnam's hamlets.[34] Ambassador Lodge questioned the big unit strategy at a planning conference in Honolulu in 1966, saying "we can beat up North Vietnamese regiments in the high plateau for the next twenty years and it will not end the war".[35] Lodge and other critics were suggesting that the United States could abandon attrition in favour of an increased emphasis on pacification. While Americans interpreted the attrition data as evidence that the United States was winning the war, subsequent events show clearly that this was not the case, and senior members of the Johnson administration and the armed forces warned about this at the time.

So, while there is evidence that more pacification-oriented strategies were mooted at the time, we must consider how these strategies, if implemented, might have changed outcomes in Vietnam. In fact, the United States did switch to a strategy that emphasized pacification and downgraded attrition following Westmoreland's recall to Washington after the Tet Offensive. Westmoreland's replacement as COMUSMACV General Creighton Abrams (pictured in Figure 4.5) gradually abandoned his forerunner's "big unit war" in the remote rural areas in favour of a beefed-up emphasis on pacification. Abrams instituted what he called a "One War" strategy with the emphasis increasingly on pacification in operations designed to "clear and hold" South Vietnamese territory rather than engage the Communist main forces.

There are reasons for thinking that this new strategy paid significant dividends and one might therefore reasonably ask why, if pacification really was more successful under Abrams, were the allies not able to terminate the Vietnam War on terms more palatable to themselves? One possible answer is that, whatever Abrams' intentions, such a strategy was forced upon him by the fact of the withdrawal of American forces from South Vietnam. Pacification proved a practical strategy to accompany a long-term withdrawal, but this could not be a war-winning strategy because allied victory

FIGURE 4.5   Westmoreland's replacement as COMUSMACV, General Creighton W. Abrams presents a Presidential Unit Citation to the South Korean Tiger Division in September 1968. Abrams gradually abandoned his forerunner's "big unit war" in favour of an increased emphasis on pacification. (U.S. National Archives and Records Administration)

was no longer possible. The US government saw the war as already lost, and the American forces available to Abrams were only going to get smaller.

Under Abrams, the increased emphasis on pacification was part of the process of withdrawal rather than a strategy designed to secure victory. The Tet Offensive and subsequent Clifford review were watershed moments in the Vietnam War, when the United States shaded over from optimism about victory to certainty of defeat. Thus, any counterfactual solution to the Vietnam conundrum from the period *before* Tet has the potential to utterly change the post-Tet landscape and thereby change the war's outcome. If the United States had not been minded to withdraw, and if pacification after Tet really did produce positive results for the allies, then a counterfactual that proposes extensive pacification before Tet offers the prospect of victory.

THE HAMLET EVALUATION SYSTEM

The data for such an assessment comes from the US Hamlet Evaluation System (HES) for the periods before and after the implementation of the new pacification programme under Abrams. HES was designed to measure the security and pacification status of all the small settlements – "hamlets" – in the South Vietnamese countryside. It had its origins in a request by McNamara, in October 1966, for the development by the CIA of a system to measure progress in the pacification effort in South Vietnam. The Agency responded with HES, which was designed to assess the relative security and pacification status of the smallest communities in Vietnam. Figure 4.6 shows the locations of hamlets in South Vietnam recorded in January 1967.

The HES system was run by CORDS with MACV responsible for its implementation. MACV District Advisors were given an allocation of hamlets to be visited monthly where they would consult with local chiefs and complete questionnaires rating the state of security and development of each hamlet. The original HES system was known as HAMLA and employed a worksheet which had 18 indicators of hamlet security and development.[36] MACV processed these scores to produce each Republic of Vietnam-controlled hamlet's comprehensive rating which ranged from "A" (friendly) down to "E" (contested) in addition to "V" (VC/NVA controlled), "P" (Planned), "X" (Abandoned) and "N" (Not Evaluated).[37] The HES data were aggregated with that from other sources in multiple reports, re-analysed and presented as evidence of progress by the US Government.

Clearly, there were problems with both the collection and the nature of the HES data. It was certainly a big exercise: according to the HES, throughout the entire war there were about 23,000

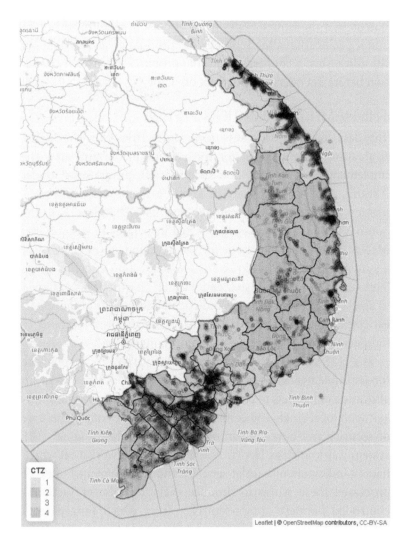

FIGURE 4.6 Map of South Vietnam. Each colour represents a different "CTZ" or Corps Tactical Zone (sometimes known as a military region). Provincial boundaries (black lines) are based on data from *The Digital Vietnam War* (Douglass 2011). Translucent black circles represent hamlets reported in January 1967. Many hamlets shift in position through time or disappear before the end of the time series, which has also been cleaned of hamlets that were reported as located

hamlets recorded in South Vietnam. The availability of computers seemed to make the manipulation of all this data practical, but the actual collection of the data proved to be a labour-intensive, time-consuming and sometimes dangerous task. The logistical complexity of the task and the clear pressure to report progress seems to have distorted the data collected, and this was not done consistently through the war. The numbers generated were based on assessments by the district advisors which were essentially subjective, which raises the issue of the extent to which assumed "objective" information could be extrapolated from them. However, William Colby, Director of the CORDS effort in South Vietnam between 1968 and 1971, felt that HES data was at least "a very handy tool", and although he acknowledged its many failings, he believed that it could still be used "to indicate trends" over time and space.[38] Douglas Kinnard found that a substantial majority of US General Officers serving in Vietnam between 1965 and 1973 were favourably disposed towards HES.[39]

It does indeed seem that the HES data act like a thermometer of the war, generally agreeing with the historical record. For example, consider Figure 4.7 where we display average security classification in HES by province for various dates. We can see the Tet Offensive, which took place between the end of January and late March 1968, reflected in the data. Tet clearly harmed the security of the entirety of South Vietnam but especially so in Thua Thien province in the north of South Vietnam, where the city of Hue was held by the Communists for nearly a month, as seen in the contrast between 1 December 1967 and 1 February 1968,

←————————————————————————————————————

outside Vietnam's borders or in the ocean! Hamlets are grouped into villages, which are in turn grouped into provinces, which are finally grouped by CTZ. (Hamlet data were obtained from the U.S. National Archives and Records Administration and (modern) background map from OpenStreetMap. The figure was produced in R using the leaflet package.)

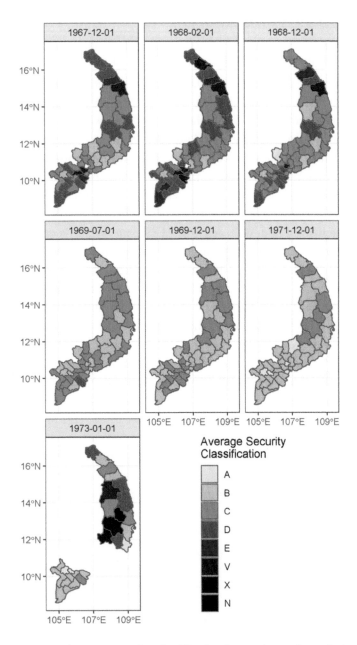

FIGURE 4.7 Average security classification in provinces through time, calculated from the median CLAS (Classification Level Indicator) of the

although this was extremely costly for the NLF.[40] Two further Communist offensives took place in 1968, the so-called "mini-Tet" in May and June 1968, and a further third offensive spasm in August and September. Overall, the Communist offensives of 1968 cost them 111,000 casualties, mostly from the NLF, inhibiting their ability to continue operating. The allies responded to the Communist offensives of 1968, and the consequent weakening of the NLF, with an Accelerated Pacification Campaign (APC) from November 1968 to January 1969, and the overall effects are reflected in the improvement in security indicated by the contrast between 1 February 1968 and 1 December 1968. Progress is slow and piecemeal through 1971, but the situation does continue to improve, and the allies appear to be winning by the end of the year. However, by 1973, the situation for "peace with honour" now looked grim and recording was actively being abandoned.

The general trend towards improved security is also detectable in Figure 4.8. This demonstrates that after the departure of Westmoreland from South Vietnam, and as the US emphasis on attrition declined and that on pacification increased, the number of hamlets under government control increased markedly while the number of contested or NLF-controlled hamlets declined. This trend accelerated from the end of 1968, when the allies implemented the APC, designed to restore losses incurred as a result of the Tet Offensive. The Tet losses were soon made good, and the allied position continued to improve – so much so that by the end of 1969, 79.1% of the (government recorded) countryside was officially under government control (rated at least C), while only

---

hamlets in a province, ordered A > B > C > D > E > V > X > N > P. Dates are chosen to demonstrate changes following major events. Tet causes a notable downturn in security, but it does not take long for these values to climb back up again. (Hamlet data were obtained from the U.S. National Archives and Records Administration and provincial boundaries data were obtained from Douglass (2011). The figure was produced in R using the ggplot2 package.)

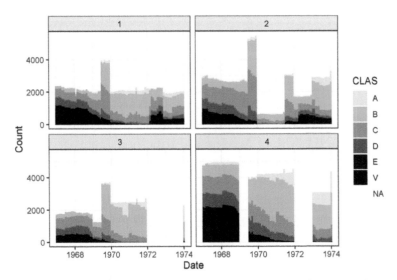

FIGURE 4.8 Security classification through time in each CTZ. Values shown are raw counts of the CLAS assigned to each hamlet in each CTZ at a given point in time, neglecting those assigned "N", "P", or "X". Values are stacked atop each other so that the total number at a given time point is the total number of hamlets at that time point, and the sections of the shaded bar correspond to the portion of hamlets with a given rating. Clearly, recording of many hamlets is spotty at best for extended periods of time, but the general trend appears to be hamlets moving into better classifications (A–C) from worse ones (D–E or V). (Hamlet data were obtained from the U.S. National Archives and Records Administration and provincial boundaries data were obtained from Douglass (2011). The figure was produced in R using the ggplot2 package.)

4.5% of hamlets were under NLF control (V) and 11.6% were contested (rated D or E) according to HES.[41] In fact, the pacification position was improving slightly right from the implementation of the HES system in January 1967 until the Tet Offensive, when the allies necessarily reduced their pacification effort in the countryside in order to fight the offensive in the major urban areas, with the result that the number of hamlets under government control declined somewhat, but the situation improved rapidly following

Westmoreland's departure. Government control remained around 80% until 1972 according to HES, when records cease for some Corps. Even amongst the records that are still present, there are upticks in 1973 and 1974, suggesting that government control at the end of HES's lifespan was around 60%.

While the above suggests trends in the classification of hamlets at the Corps Tactical Zone (CTZ) and provincial levels that match the general historiography of the war, the trends also match some specific details of the historiography. For example, units from the Republic of Korea had a reputation during and after the war for poor behaviour towards the local Vietnamese population.[42] This is reflected in the HES data when combined with the identity and locations of units in the theatre (recorded from Southeast Asia Friendly Forces File [SEAFA]). Although most of the association between units and hamlet security scores is weak, some trends do emerge. From 1967 to 1969, units from the Republic of Korea were associated with hamlets that generally had security scores that were becoming worse, not better, over time. Even hamlets that were improving in security score saw less improvement than would otherwise be expected (on average).

Similarly, the general trends towards improved security seen across South Vietnam from 1968 are reflected in Figures 4.7 and 4.8 for 1CTZ, which was the responsibility of the US Marine Corps' III Marine Amphibious Force. These positive security figures for 1CTZ probably reflect the added emphasis that the Marines placed on pacification compared with their US Army brethren. Senior Marine commanders were among the most outspoken critics of Westmoreland's attritional strategy, and the Marines' approach included combining Marine infantry squads, including a Navy corpsman, with South Vietnamese territorial militia platoons into "Combined Action Platoons" (CAPs). These Marines would live in the hamlets alongside their South Vietnamese colleagues, providing additional local security, seeking out the NLF infrastructure and training the South Vietnamese so that they could eventually take over full security responsibility for their locality. According to HES, all five provinces of 1CTZ were either contested or under NLF control at the end of 1967 (see Figure 4.7, 1 December 1967) and security conditions deteriorated further as a result of the Tet Offensive, but 1 July 1969 shows improvement across the board in 1CTZ, by which time the entire CTZ was judged relatively secure. Conditions continued to improve to the end of 1971, by which time the Marines had withdrawn most of their combat forces.

Westmoreland, however, thought the CAPs an inefficient dispersal of Marine strength in the face of the Communist main force threat.[43] In practice, other demands on Marine manpower meant that only a fraction of Marine strength was deployed in this manner, and no CAPs ever achieved the desired end state allowing them to withdraw, secure in the knowledge that their localities were adequately protected.[44]

PICKING UP THE PIECES

If the objective of counter-insurgency operations is the defeat of the insurgent infrastructure, the reduction of the area controlled by the insurgents and the consequent increase in that area controlled by the government, then the data suggest that the allies were making progress – winning the war – following the application of an increasing emphasis on pacification. The problem, of course, for the Americans, was that while they might have hit upon – or at least been forced into – a winning strategy, they had already accepted defeat in practice. They reduced their influence on South Vietnamese strategy and ultimately withdrew effective support for the South Vietnamese in the face of the North Vietnamese conventional offensive of 1975.

General Abrams might therefore be said to have been winning the war after Tet and only ultimately did not prevail because the US Government had already decided the war was lost. We can speculate that earlier American concentration on pacification might have produced victory for the allies or at least allowed them to do rather better with greater prospects for a negotiated settlement. The HES data suggest that this scenario promised a greater likelihood of success on the ground in South Vietnam and an approach to the war that would have been more tolerable to the American people in terms of troop levels, combat intensity and effect on world opinion, rendering the war politically more sustainable for the American public. Even then we cannot know for certain that such a strategy would have resulted in an allied victory, but the evidence of the HES data allows us to envisage that such a strategy might have allowed the United States to have sustained its war effort for longer. It has often been said that in order for the Communists to win the Vietnam War they needed only not to lose it, but it could equally be said that for the Republic of South Vietnam to win it needed only to survive. Indeed, while the Communists were not winning, they were effectively losing because of the continued survival of South Vietnam. An earlier increased emphasis on its pacification seems to offer the promise of the extended survival of the South Vietnamese republic as a result of continued US engagement.

That said, we must acknowledge deficiencies within HES itself, as well as the old (even at the time of the Vietnam War) adage "garbage in, garbage out". HES by itself appears to be consistent with our understanding of the Vietnam War: when people say security is increasing, HES reports increases in security. However, the same people who were saying that security was increasing were also generally consumers or providers of HES data, creating a potentially tendentious information loop.

So, if progress can be measured in terms of pacification results, does it follow that increased emphasis on pacification might produce still more progress? Part of the difficulty in arguing along these lines is that it is hard to measure from the data how and when such a change might have occurred, but we can apply the context to make an educated guess. As we have seen, such a transformation did occur, to some extent, with the replacement of Westmoreland by General Abrams and the introduction of accelerated pacification. The fact is, however, that attrition was always going to be a major part of US strategy while Westmoreland remained as COMUSMACV. We can speculate that the emphasis on attrition would have been reduced in favour of enhanced pacification efforts earlier if Abrams had succeeded Westmoreland as COMUSMACV earlier. Indeed, Westmoreland's term of office as COMUSMACV, at more than four years, was unusually long for an American theatre commander. When Abrams became Deputy COMUSMACV in May 1967, it was on the understanding that he would become COMUSMACV in due course, but he did not succeed Westmoreland until June 1968. Lewis Sorley has suggested that there is evidence to indicate that Abrams was supposed to succeed Westmoreland earlier, within perhaps two or three months of his arrival in Vietnam, but this timescale slipped for reasons that are unclear.[45]

Given that Westmoreland's emphasis on attrition was the subject of criticism by military subordinates, superiors and other commentators long before Abrams arrived in South Vietnam, we might push our counterfactual emphasis on pacification back still further to 1966 or even 1965 when General Johnson, a critic of Westmoreland's strategy, proposed his PROVN plan, or perhaps even to 1964, when Seymour Deitchman proposed a strategy for guerrilla insurgencies – as he understood the Vietnam war to be – in his book *Limited War and American Defence Policy* (1964).

As we have seen, Deitchman showed in his 1962 paper that the most effective way to respond to a guerrilla insurgency was to adopt the same techniques as the guerrillas. This is precisely what the United States Special Forces set out to do in the early days of US intervention in the Vietnam War. This was not a population-centric pacification strategy, but it was certainly not an attrition strategy either. Deitchman fleshed out the details of a holistic counter-insurgency strategy in *Limited War*, where the main aspects of Deitchman's formula were as follows: (1) a hearts-and-minds campaign to secure the support of the population; (2) a campaign to ensure the security of the population from guerrilla agitprop and terrorism, and to cut the guerrillas off from their base of support in the population; and (3) an offensive against the guerrillas conducted along the lines of the guerrillas' own campaign, by small counter-insurgency units deployed permanently in guerrilla territory. Deitchman believed that such a campaign must be the responsibility of the defending nation and the best role for the United States would be the provision of advice and support rather than direct military intervention.[46] Deitchman remained convinced that his proposals constituted a solution to the Vietnam problem as it stood in 1964: "I had worked out a very simply stated strategy for counter-insurgency: you have to defend the population, you have to go after the insurgents, and you have to change the conditions that induce the population to help the insurgents" but he also accepted that "although the strategy is simply stated it can be fiendishly difficult to carry out".[47]

In proposing our counterfactual scenario of earlier intensified pacification, we have to be mindful of the risk that if we change one variable – in this case substituting pacification for attrition – this might have knock-on effects for which we cannot adjust. For example, the greater manpower needed for pacification might require moving closer to the theoretical ten-to-one ratio of security forces to insurgents of traditional COIN doctrine. Neither should we discount the possibility that the absence of the big-unit search and destroy sweeps might open up South Vietnam to greater threat from the PAVN and NLF main force units, which the US Army would not now be able to keep off balance.

However, if it is the case that the Communist main forces were almost always engaged by Westmoreland's search and destroy operations only when they chose to be, then Westmoreland was not keeping the Communist main forces off balance at all, and increased pacification does not increase the threat from Communist main force units. Such a threat continued to exist, however, as evidenced by the occurrence of the Tet Offensive in 1968, and subsequent major Communist Offensives in 1972 and 1975. If our increased pacification counterfactual had been implemented, it might actually have had the effect of taking the brakes off further Communist offensives, or even have forced their pace, out of Communist frustration with their lack of progress in the countryside. In the real Tet Offensive of 1968, the effect had been devastating on the allied cause because the Johnson administration's relentless boosterism rendered the American public psychologically unprepared for a major Communist offensive, even if it was handily defeated by the allies – as was, in fact, the case, leading to the development of the credibility gap between what the Johnson administration said and what the public believed, and a consequent collapse of political will to continue the war. "In a sense, 'the system worked'[48] in that conscious decisions to prevent the South collapsing were taken, and it was only once the United States lost the motivation to continue intervening that the North was able to win its decisive victory in 1975".[49] Indeed, it might reasonably be argued that the stimulation of major Communist offensives was all to the good as far as the allied cause was concerned because it would allow the United States to employ its firepower advantage against the Communists – just as it did in Tet 1968, and just as the US armed forces always argued they wanted to do, provided the American public were adequately prepared for major enemy offensives. This preparation would be taken as a state of readiness on the ground that would also help align all CTZ's strategic priorities with the counterfactual COMUSMACV's.

In 1968, Assistant Secretary of Defence Paul Warnke of the Department's Office of International Security Affairs (ISA) proposed such a state of readiness in a memo to new Secretary of Defence Clifford in which he argued against Westmoreland's post-Tet request for 206,000 additional troops and proposed an alternative MACV strategy that emphasized counter-insurgency over attrition. Similar to the enclave strategy proposed earlier in the war, Warnke proposed that US forces concentrate on protecting the population from bases in the heavily populated coastal region of South Vietnam, or just beyond it, on the "demographic frontier". The large search-and-destroy operations deep in the interior would cease in favour of "spoiling raids, long-range reconnaissance patrols and, when appropriate targets are located, search-and-destroy operations into the enemy's zone of movement in the unpopulated areas between the demographic and political frontiers. They would be available as a quick reaction force to support the RVNAF when it [sic] was attacked within the populated areas". Warnke thought that this would actually release more forces for security tasks and forces would still be available to cope with future Communist main force offensives.[50]

CONCLUSION

Our counterfactual has two components: increased pacification and a declared readiness to profit from precipitate Communist offensives such as occurred in 1972, when the ARVN, supplemented by US airpower, proved sufficient to defeat the Communist offensive without the presence of American ground troops. The final allied objective would still have been negotiations. The best we can say is that if our counterfactual had become fact, South Vietnam might have survived, for longer, allowing a more successful nation-building effort, and or a negotiated solution under more favourable circumstances than in 1973.

In practice, of course, the US digital windfall in Vietnam clearly did not guarantee victory, and perhaps the main lesson of the war is that access to copious amounts of data is not inherently useful. More important is the quality of the data, its fitness for purpose and the facility to interpret it and act upon it. Poorly used data, as much as any other tool, can have a detrimental effect on a campaign. Just as one should not fire bullets for the sake of reaching a certain number of shots fired, one should not collect data for its own sake. The body count data seem to have been a case in point.

We have already established that both the body count and HES data were problematic, but both could be said to indicate general trends about Communist casualties and the pacification of the South Vietnamese countryside. It is perhaps even the case that body count was little better or worse than HES in terms of quality. What made body count less useful than HES was that it was manipulated to show progress where perhaps there was little or none, and more fundamentally body count was objectively a poor metric for measuring allied progress. While, on the face of it, measuring the number of enemy casualties sounds like it should tell us something useful about progress, it could not reveal how close the Communists were to achieving their objectives in South Vietnam or how close they were to defeat, no matter how good the quality of the data. It really would not matter if body count revealed that the Communists were going to run out of troops in 1976 if they had won the war by 1975!

Instead, data need to be collected with a plan and a purpose and there are clear reasons why methodology is so closely vetted today. With these caveats, we can say that the HES data are supportive of a counterfactual based on an abandonment of attrition in favour of enhanced pacification.

More broadly, Vietnam and MACV's quest for data demonstrates the problem of having so much data that it blinds you – just as the Battle of Britain demonstrates the advantages of an information-filtering system which gives you the right data for the job.[51] This is not a uniquely Western capitalist problem either; as we will see in the coming chapter, the USSR was just as susceptible. In examining the USSR's struggle with data, we will be able to explore similar sorts of counterfactual mindsets to those the United States imputed to North Vietnam, and how a mind-game emerged that could have led to nuclear war.

NOTES

1. Thayer, *War Without Fronts.*
2. McNamara, *In Retrospect,* pp237–238.
3. See Daddis, *No Sure Victory* & Connable, *Embracing the Fog of War,* pp95–151.
4. Andrade, "Westmoreland Was Right".
5. Sorley, *Westmoreland,* pp103–104.
6. Sorley, *Westmoreland,* p123.
7. Daddis, "Choosing Progress".
8. Clifford, "A Viet Nam Reappraisal".
9. Hess, *Vietnam,* pp12–19.

10. For examples, see Vietnam veteran Harry G. Summers Jr. argued in favour of an incursion into Laos in *On Strategy*; Admiral U.S.G. Sharp, who was CINCPAC and responsible for the US air war over North Vietnam until 1968 – this was not part of COMUSMACV's remit – argued that political limitations thwarted the decisive use of US air and naval power in Vietnam in *Strategy for Defeat;* and Vietnam veteran Andrew F. Krepinevich Jr. argued that slavish adherence to what he called the "Army concept": "a focus on mid-intensity, or conventional, war and a reliance on high volumes of firepower to minimise casualties …." prevented the US Army from paying adequate attention to the pacification of the most populous regions of South Vietnam in *The Army and Vietnam*, pp4–5.
11. Shultz, "Breaking the will of the enemy".
12. See Dror Yuravlivker, "Peace Without Conquest".
13. For a discussion of this point, see Connable et al., *Will to Fight*.
14. Thayer, *War Without Fronts*, pp28–29.
15. Graham would go on to be Deputy Director of the CIA and Director of the Defence Intelligence Agency. Wirtz, "Intelligence Please?", p245.
16. Wirtz, "Intelligence Please?", pp248–249.
17. Thayer, *War Without Fronts*, pp30–31.
18. CBS, *The Uncounted Enemy: A Vietnam Deception* (23 January 1982), *https://www.youtube.com/watch? v=PFXTCfY5qME, Accessed 30 June 2022. Westmoreland* finally settled out of court in 1985.
19. McNamara, *In Retrospect*, pp239–242.
20. Wirtz, "Intelligence Please?" pp249–251.
21. Thayer, *War Without Fronts*, pp89–90.
22. McNamara, *In Retrospect*, p263.
23. Thayer, *War Without Fronts*, p90.
24. Thayer, *War Without Fronts*, pp95–96.
25. Deitchman, *Lanchester Model of Guerrilla Warfare*.
26. Of course, the allies will then also have to accept higher casualties, which in Deitchman's model scale as insurgent numbers, or equivalently (which is more useful since insurgent numbers are usually unknown!) as the square root of allied numbers. This conclusion was arrived at empirically from historical data by Goode, without his making a connection with Deitchman, whose theoretical prediction he was unknowingly confirming. Goode, S.M. A historical basis for force requirements. Parameters, Winter 2009–2010, p45–67.

27. The allied manpower advantage fell away significantly if one considers the large number of non-combatants in the allied order of battle, particularly in the US Army. According to Thayer, while the allies always outnumbered the Communists by at least 3.5–1 and by nearly 6–1 between 1969 and 1971, what he called the "foxhole advantage", the actual number of combat troops available for operations, hovered around 1.6–1 between 1965 and 1971. If we consider the number of troops available for offensive operations, the ratio falls even further to around numerical parity, with the Communists even enjoying the advantage at times. It is true that the allies generally enjoyed considerable advantages in terms of firepower, manoeuvrability and combat support, but the Communists proved adept at mitigating these advantages by using ambush tactics and in the case of firepower the Communist actually held some advantages at the small unit level. Thayer, *War Without Fronts*, pp92–95.

28. Enthoven and Smith, *How Much is Enough?* p298.

29. Although the review certainly took place, no records of its deliberations were compiled. We have only the reflections of the participants to go on. Clifford, "A Viet Nam Reappraisal". pp609–612.

30. For the view that PROVN represented an outright rejection of Westmoreland's strategy, see Sorley, *Westmoreland*, p104 and Sorley, *Honourable Warrior,* pp227–241; ODCS military Operations, Programme for Pacification, p5; Andrade, "Westmoreland Was Right", p158.

31. Sharp and Westmoreland, *Report*, p292.

32. Mueller, "The Search for the 'Breaking Point'", p517.

33. Krepinevich, *The Army and Vietnam*, pp262–263.

34. Document 294, Ambassador to Vietnam (Lodge) to President Johnson, Saigon, 7 November 1966, *FRUS 1964–1978, Vol. IV, Vietnam 1966*; Buzzanco, *Masters of War*, pp203, 246–251, 260–261; Sorley, "To Change a War", pp93–109.

35. Quoted in Sorley, *Westmoreland*, p104.

36. HES received multiple updates during the time it was used. The major updates were HAMLA, HES70 and HES71. National Archives and Records Administration, HES71 and HAMLA Technical Documentation, p13, 2019.

37. Young, "Computing War Narratives", p56 and National Archives and Records Administration, HES71 and HAMLA Technical Documentation, p13, 2019.

38. Connable, *Embracing the Fog of War*, p113.

39. A strong majority, 77%, of US General Officers who responded to a questionnaire from Kinnard felt that HES was either, "A good way

to measure progress in pacification", or "Had weaknesses but was about as good as could be devised". In contrast, 19% felt that it was "not a valid way to measure progress in pacification", while 4% did not answer the question. Of 173 US Army Generals who served in Vietnam between 1965 and 1973, 64% completed the questionnaire. Kinnard, *The War Mangers*, p108.

40. John T. McCormick has also found that a "positive sign for the [HAMLA] data's ability to map the progress of the war is that during the period associated with the Tet Offensive (January–March 1968) we can see an increase in VC held territory in and around when the fighting was most intense". McCormick concludes: "though far from scientifically conclusive, this [his] study has broadly shown that there is analytic potential and quantitative validity in the HES dataset". "The Hamlet Evaluation System – Reevaluated", (10 May 2021), https://storymaps.arcgis.com/stories/1c1bb536ec494f69815ab27feef47255, accessed 4 January 2023.

41. The remaining approximately 4.9% fall under the category of planned, abandoned or not evaluated.

42. The US CORDS reported dissatisfaction with the pacification performance of South Korean forces during the war, and subsequently the ROK corps in Vietnam has become associated with multiple atrocities, possibly resulting in the deaths of several thousand South Vietnamese civilians. Larsen and Collins, *Vietnam Studies, Allied Participation in Vietnam*, pp158–159. For some examples of articles relating to ROK atrocities in the Vietnam War see Dien Luong, "It's Time for South Korea to Acknowledge its Atrocities in Vietnam"; Julian Ryall and The Korean Times, "Moon Jae-in's Administration Faces Call to Investigate War Crimes Amid Rising Awareness of South Korea's Atrocities in Vietnam"; Heonik Kwon, "Anatomy of US and South Korean Massacres in the Vietnamese Year of the Monkey, 1968".

43. Hennessy, *Strategy in Vietnam: The Marines and Revolutionary Warfare in I Corps, 1965-1972*, pp75–77 and William C. Westmoreland, *A Soldier Reports*, pp214–216.

44. Andrade, "Westmoreland Was Right", pp158–159. Kalyvas and Kocher, in "The Dynamics of Violence in Vietnam", also affirm the ability of the HES to indicate general pacification trends, confirming an anticipated correlation of the incidence of violence in the South Vietnamese countryside with HES hamlet security classifications. In their study, Kalyvas and Kocher differentiate between terroristic "selective" violence as practised by the communist NLF and "indiscriminate" violence as inflicted by allied firepower and

find that NLF selective violence has its highest incidence in areas predominately, but not fully, controlled by the insurgents and allied indiscriminate violence had a higher incidence in areas fully controlled by the insurgents.

45. Sorley, *Honourable Warrior*, pp270–273; Sorley, *Thunderbolt*, pp194–195.
46. Deitchman, *Limited War*, p40.
47. Deitchman, Sheldon and Wong, Interview, p71.
48. Gelb and Betts, *The Irony of Vietnam*.
49. Gelb and Betts, *The Irony of Vietnam*; McCann, "Killing is Our Business", p501.
50. Krepinevich, *The Army and Vietnam*, pp242–243.
51. Holwell and Checkland, *An Information System Won the War*.

CHAPTER 5

The Road to Able Archer

Counterfactual Reasoning and the Dangerous History of Nuclear Deterrence 1945–1983

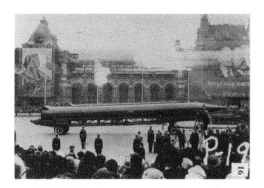

FIGURE 5.1 The SS4 "Sandal" medium range ballistic missile (MRBM) pictured here and its larger cousin the SS5 "Skean" intermediate range ballistic missile (IRBM) were modern and effective weapons. Neither had been designed to reach the United States from the USSR, but American U2 reconnaissance flights and "Corona" satellite imagery revealed by 1961 that Soviets had no effective strategic deterrent. Khrushchev's decision to place these missiles in Cuba remedied this deficiency but destabilized the deterrent balance with the United States catastrophically in the opposite direction. (Dino A. Brugioni Collection, National Security Archive, Washington, DC)

DOI: 10.1201/9780429488405-5

HISTORICAL CONTEXT

By 1983 the nuclear arms race had been going on for over 30 years, coming closest to Armageddon most famously in the Cuban missile crisis of 1962. A lesser-known crisis is that of 1983, when a NATO command exercise, Able Archer, may have caused the USSR, on the basis of faultily gathered intelligence data, to believe that a NATO nuclear strike was imminent – and to consider pre-empting it. It would be easy to label this as paranoia – but this would merely hide a symmetrical problem in which each side, moving in a world of strategic games, nevertheless totally failed to appreciate the other's perceptions and position, thereby placing themselves in dangerously counterfactual mindsets.

In the earlier chapters of this book, we examined relatively simple data and introduced some new ways to use it to quantify probabilities. This allowed us to introduce some restrained counterfactuals and re-think commanders' actions. Even these methods required modern computing power – recall that MCMC and bootstrap calculations alike involved hundreds of thousands of iterations. With Vietnam, in the previous chapter, came the arrival of what we might now call "big data" – but, even with the help of modern computers, this yielded no simple lessons; with the computers available in the 1960s, the likelihood of extracting meaningful truth, at least in a useful time scale, was very low indeed.

However, Vietnam represented a departure in a fundamental way. It was the first time that a government used big data in war to shape national policy. The use of operational research in both world wars to solve specific problems pointed in this direction, and in Vietnam the titanic effort in data collection and processing produced results that were trusted by government. These results have been discredited in subsequent historiography as a viable guide to policy, and policy makers themselves stand accused conversely of failing to be guided by the data, instead interpreting it falsely to justify existing policy assumptions. A surfeit of information and poor analysis of it can result in a fundamentally incorrect perception of operational reality. Further, too much data, compounded by too much belief that it must be telling you something, can too easily downplay the value of expertise, experience and judgement. Giving the data primacy can deny human agency in an exact reversal of the intended effect of using it to disperse the fog of war and enable decision-making. Ironically, as the previous chapter demonstrated, the US government having used data to justify falsely optimistic assumptions then became gripped by pessimism as to the outcome of the war and ignored data that was telling them, albeit with caveats, that the post-Tet Vietnamization policy was working as intended.

However, Vietnam was not the only area in which the United States used mathematical modelling to aid decision-making and instead became a prisoner of its *a priori* assumptions. The burgeoning nuclear arms race following the detonation the first Soviet A-bomb "Joe 1" in 1949 demanded accurate assessment of both Soviet nuclear strength and the tenets of its nuclear decision-making process. The new science of "game theory" suggested itself as an excellent tool for rationalizing the problems bedevilling the nuclear decision-making process. In a sense, however, the United States faced the opposite problem to Vietnam, a dearth of data rather than an excess of it. It did, however, make exactly the same mistake, basing assessments of Soviet nuclear capability and decision-making on fear and fanciful assumptions while believing that it was guided by the firm hand of scientific method. The worst effects of this were mitigated, in that the Soviets, unhampered by the information desert that their secretive ways created for the Americans, understood the situation perfectly for most of the Cold War and understood in particular that they were in a position of dangerous material inferiority. Their success in persuading the Americans to adopt the policy of Shakespeare's Dauphin in Henry V – that in cases of defence it is best to weigh the enemy more mighty than he seems – was actually a stabilizing factor, given that the Soviets had no interest in attacking a superior opponent.

The danger, however, would come in the late 1970s, when the Soviets had achieved parity but lost their grasp of the US policy-making process. As we shall see, when both sides adopted a false or "counterfactual" assessment of their adversaries' capabilities and aggressive intent, the calculations of game theory led not to stability but to appalling danger. This was revealed in the *Able Archer* crisis of 1983. An academic debate has developed as to the extent to which this NATO command post exercise did or did not bring the world to the brink of nuclear war. In our view, this debate is irrelevant to the key outcome of the crisis, which was the culminating event of a long historical process. Our contention is that the instability caused by both sides applying a counterfactual perception of their opponent in game theory made such a crisis dangerously likely. Had it not been Able Archer, it would have been something else. The key point is that the long-feared miscalculation in a nuclear crisis arrived in a form where swift de-escalation was possible. Even more importantly, the scale of the misperception became clear to both sides very shortly afterwards, with wholly positive effects for their subsequent relationship.

This chapter, then, is all about human agency – and above all about the dangers of misconceiving others' actions and intentions. The crucial failure is of the theory of collective mind – "We think …, we think they think …, we think they think we think …" and its mind-bending infinite regress.[1] The context is the long history of deterrence theory, and its culmination in the paradox of adversarial states planning to avoid nuclear war yet, in Able Archer, unwittingly coming close to bringing it about. This episode reminds us forcefully that historical actors were fully capable of placing themselves in counterfactual mindscapes which might have resulted in disaster, even as they sought to envisage probable futures with scientific precision to provide early warning and thereby avoid such a fate. This tendency of historical actors to construct postulated worlds to inform decision-making is often overlooked in academic history, which disregards such necessary conceptualisation of probable scenarios and deals only with outcomes, armed with knowledge of the actors' actual future.

Although the story includes the collection and use of bad data, the main mathematical context is "game theory", which Cold War actors dealing with deterrence used as a foundation for decision-making.[2] Matters relating to nuclear deterrence are still pressing despite the end of the formal Cold War and present-day constructors of models aimed at preventing nuclear exchanges are highly conscious of the need for effective analysis of past episodes such as *Able Archer*. Thus, "more than one may assume, explorations of the Cold War strategic interactions offer valuable insights that can inform the current deterrence research program and address its theoretical voids".[3] The principal danger is of the scientific fallacy known as the "streetlight effect", the use of quantitative models because they are available rather than because they are correct.[4] When faced with perhaps the most subtle of tasks, the analysis of human understanding and communication, the models proved inadequate.

ABLE ARCHER AND ITS CONTEXT

November 1983 was a time of heightened Cold War tension and heated superpower rhetoric as the so-called Second Cold War reached a peak of intensity. It has been alleged that *Able Archer*, a NATO "command post" exercise in Western Europe designed to test communications in the event of nuclear war, was interpreted by the Soviet Union as the first stage of an actual attack.[5] The Soviet leadership thus braced itself for the thermonuclear blow between 7 and 11 November 1983 and prepared a full and possibly pre-emptive response. If true, this incident would have placed the world "on the precipice of a catastrophe"[6] – certainly comparable to the Cuban Missile Crisis of 1962, and eclipsing other nuclear crises such as the US nuclear alert during the Yom Kippur War of 1973. However, the 1983 crisis differed in character from these earlier episodes in important and alarming respects. Chief among these was that in 1983 the West might have generated a nuclear exchange completely unwittingly, as NATO planners and politicians lacked a clear understanding of their Soviet opponent's perceptions both of themselves (specifically that they were vulnerable to surprise NATO attack) and of NATO's intentions (which were never to make such an attack, for NATO's exercise was simply that). The absence of an accurate view of motives and fears on either side created basic problems for deterrence.

In other respects, however, the crisis was entirely typical. The Able Archer crisis arrived suddenly and was over quickly, but it had very deep roots in nearly 40 years of planning for nuclear war on both sides of the Iron Curtain. In studying it, we have to consider the calculated futures of the past, of chronologically successive and distinct periods in strategic deterrence, each capable of interpretation very broadly in terms of game theory, which was particularly prominent in the American analytical process. Mutual misapprehension in 1983 continued a long tradition of misunderstanding which had always created catastrophic potential for war, based consciously and unconsciously on game theory and its erroneous assumption that rational actors were guided by accurate information. In this way, they stumbled towards a war that neither had willed.

The first phase of the Cold War, which began effectively with the Berlin Blockade of 1948, was a period of actual and then effective American strategic monopoly lasting until the Cuban missile crisis of October 1962. In this period, both superpowers were nuclear-armed, but the USSR had no effective means of attacking the continental United States with nuclear weapons, whereas the United States became fully capable of assaulting the USSR with impunity. This was followed by the era of Mutual Assured Destruction (MAD), in which both superpowers could destroy the other in any circumstances. The third phase, from the collapse of *détente* in the mid-1970s and culminating in the Able Archer crisis, was a period in which MAD was questioned and the possibility of a "winnable nuclear war" was again raised. In each of these periods, the superpowers attempted to model the likely behaviour of the other in order to formulate deterrent strategies. The problem, of course, was that the superpowers were not necessarily aware of the true situation. Each sought to conceal its nuclear capabilities from the other and to actively mislead its potential adversary. The game theory of the period, as we shall see, assumes that each player has perfect knowledge of the other – that both sides know all the options and payoffs. The problem was that the adversaries were, in truth, playing different games, and the other did not know it. Sometimes the deception was deliberate, but sometimes it was unwitting and sometimes the two were present at the same time. Counterfactual perceptions of the nuclear balance developed, which became real to the extent that they were believed and decisions made accordingly, creating great danger and resulting, despite more than three decades of Cold War crisis management experience, in the near-war of 1983.

PHASE 1: AMERICAN STRATEGIC MONOPOLY AND SOVIET DETERRENCE

The advent of the atomic bomb in 1945 created an entirely new situation in strategy that few had any idea how to address. Given the highly technical nature of the problem it is perhaps not surprising that American academia established a role in conceptualising the new situation and shaping developments. This role was embodied ultimately in the RAND corporation, originally an Air Force funded think-tank housed in the Douglas Aircraft Corporation which became an independent institution and attracted largely mathematical and scientific academics of great promise or reputation, predominantly economists or mathematicians, who were interested in strategic issues of the nuclear age. They hoped to bring the fears and uncertainties of the nuclear age under some sort of rational control. Their tools were the new techniques of game theory and systems analysis, which would help planners to understand, respectively, the burgeoning capabilities of the new weapons and the mysterious motives of Soviet leaders who might use them against the United States.

Churchill's description of Russia as a riddle wrapped up in a mystery inside an enigma perfectly summed up the problems faced by the United States in divining Soviet capabilities and intentions. Little hard information was available. The controlled nature of Soviet society, its geographical isolation and the unlimited oversight of the state rendered the traditional tools of espionage ineffective. The vast size of the USSR, larger than the United States and China put together, made visual reconnaissance difficult in an age before satellite surveillance. Intelligence was largely a matter of guesswork before U2 flights began in the late 1950s. In such circumstances "an attenuated flow of information was therefore inevitable",[7] and it seemed wisest to assume the hostility of the Soviets, including a basic intent to attack the United States, but with the caveat that the Soviet leadership was rational and would only indulge in naked aggression if it was sure of victory. In the absence of knowledge, the assumption of rationality provided analytical possibilities. The American mathematician John von Neumann, who was closely involved with RAND, devised the techniques of game theory in the context of economic analysis to predict the behaviour of consumers, but its wider applications soon suggested themselves. RAND analysts were much interested in the military *Kriegsspiel* tradition of training for future war through wargaming, and it has been argued that game theory was "the twentieth century's kriegsspiel".[8] The most famous example is the Prisoner's Dilemma game, so named because in the classic example two notional apprehended criminals are held in separate cells and each is given the option to inform on their colleague for reward. If one informs but the other remains silent, the informer goes free but his partner is imprisoned for, say, three years. If they inform on each other, they each get two years in jail, but if both remain silent they each go to prison for one year. The game centres on the payoffs or penalties for each depending on how they act and the possibility of a rational or even mutually beneficial course of action based on the criminals' separate deductions. The precise prison terms don't really matter; what is important is the ordering of outcomes, and positive numbers are easier to deal with, so we take the payoffs to

Second Suspect

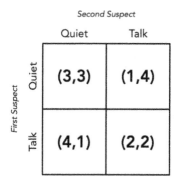

FIGURE 5.2   Prisoner's Dilemma (Pd).

be 1 (worst) through to 4 (best).[9] The options are usually portrayed in the form of a matrix as in Figure 5.2, in which the notation brackets (first suspect's payoff, second suspect's payoff). In the text, we also write (*first suspect's strategy, second suspect's strategy*).

There is no obviously correct course of action. Game theory uses three ideas to think about the possibilities. A strategy is **dominant** if it is your best strategy for each possible choice by your opponent – that is, it is best for you whatever the other player chooses. It is almost like not playing the game at all – you don't need to care what the other guy does. Here, the dominant strategy is to *talk*: if the other player is *quiet*, I am better off *talk*ing; likewise if the other player *talk*s, I am better off *talk*ing. (*Talk, Talk*) is also what is called a **Nash equilibrium**: no one can unilaterally do better by choosing to do something different. And yet there is clearly something wrong with all of this; the Nash equilibrium makes one feel trapped. If the players cooperated, they could both do better, and (*Quiet, Quiet*) is what is called **Pareto-optimal**: there is no other state in which someone does better and no one does worse. (*Talk, Talk*) is clearly not Pareto-optimal – although (*Talk, Quiet*) and (*Quiet, Talk*) are. To achieve (*Quiet, Quiet*) requires good communication rather than supposition and unilateral action – and trust instead of selfishness.

The example of Prisoner's Dilemma was thought to become relevant to the conceptualization of war if the prisoners are replaced by the two superpowers faced with the decision of whether or not to launch a nuclear strike, as it modelled "conflict among rational but distrusting beings".[10] This then becomes a game of deterrence. The key point is that the superpowers, like the prisoners, had to make decisions based on deductions without complete information.[11] In these circumstances, for each superpower, the other became what it was perceived to be, not necessarily what it really was. Likewise, in the prisoner game, each player is effectively conceptualising two colleagues: one who can be trusted and one who cannot. In both cases a decision must be made on an effective choice between one counterpart who is real, and one who is not (and thus purely counterfactual). For the decision-maker, each has a probability of being real until the truth is revealed and the fictional counterpart evaporates. Knowing which is real can only rely on rational judgement or prior knowledge, preferably the latter. However, understanding the opponent was not easy – "It may not seem reasonable to you but it probably does to him. Rationality, after all, can come in many different forms."[12] In the 1950s, the Soviet leaders were much more adept than their US counterparts in understanding the opposing rationale and, to put the matter crudely, a remarkable situation came into being that the United States was bluffed into believing in the phantom, and the real Soviet player won the game.

Prisoner's Dilemma, used for nuclear war with *Talk* replaced by *Strike (first)* and *Quiet* replaced by *Don't (strike)*, makes a rather strange assumption, that to *strike* first is better than that both sides *don't*. An alternative is the game known as Stag Hunt, after a parable told by Rousseau, in which two hunters must cooperate to catch a *Stag*. Either may choose to defect, and catch (perhaps only a share of) a much smaller *Hare*, as shown in Figure 5.3.

The only difference from Prisoner's Dilemma (with *Stag* replacing *Quiet* or *Don't*, and *Hare* replacing *Talk* or *Strike*) is that both sides have swapped their two most favoured outcomes. However, the game-theoretic optimum is now rather different, for *Hare* is no longer dominant – and so is not obviously the best choice if one prefers

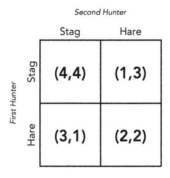

FIGURE 5.3   Stag Hunt (Sh).

not to think at all – and (*Stag, Stag*) is now another Nash equilibrium (as well as remaining Pareto-optimal). (*Stag, Stag*), or (*Don't,
Don't*), is called a "payoff equilibrium" – it is the better, clearly.
(*Hare, Hare*), or (*Strike, Strike*), is a "risk equilibrium" – roughly
speaking, it is preferred if there is uncertainty about the other side's
state or intentions. In nuclear war, Stag Hunt is usually the correct
model, with both sides preferring to avoid nuclear war, unless one
side has first-strike capability (the capacity to decapitate or destroy
its opponent with something close to impunity) and the desire to
use it, in which case its favourite two outcomes swap order, and the
game becomes the Asymmetric Dilemma of Figure 5.4.

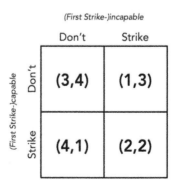

FIGURE 5.4   Asymmetric dilemma (PdSh).

In Asymmetric Dilemma, as in Prisoner's Dilemma, (*Strike, Strike*) is the sole Nash equilibrium, while (*Don't, Don't*) is Pareto-optimal. Most worryingly, (*Strike, Don't*) is dominant for the first-strike-capable (only). If there is no talk, no trust and the incapable side believes that the capable side wishes to absent itself from the game, then it believes that a strike will come – and, it might be argued, in keeping with the Nash equilibrium, should itself strike if and when it becomes capable of doing so.

By now, with games proliferating, we need some organizing principle, some "periodic table" for games, and indeed such a classification exists (Figure 5.5). There are 78 different games, organized into a 12 × 12 square by the two sides' orderings of the four outcomes, with symmetric games (in which both sides use the same ordering of outcomes, as in **Pd** and **Sh,** written in the table as **PdPd** and **ShSh**) along the (top-)left-to-(bottom-)right diagonal.[13] Asymmetric games are classified by combining the two-letter codes, so that Asymmetric Dilemma, which has **Pd**-ordered outcomes for the first-strike-capable and **Sh**-ordered outcomes for the incapable, is **PdSh,** which you can observe in the middle of the square, just off the diagonal. Games are adjacent if one side's swapping two payoffs switches from one game to the other – so **Pd** is adjacent to **PdSh**, **PdSh** to **Sh**, and **Pd** is two places (or one diagonal place) from **Sh**. Now, 78 may seem a lot – but clearly swapping of 4 and 3 matters rather more than swapping the less favoured outcomes, so we place **Sh** and **Pd** at the centre, making the distinction between **Sh** and **Pd** literally central to the argument.

Then stepping a few squares down from **Sh,** we reach games with names such as Peace (**Pc**), Harmony (**Ha**) and Concord (**Nc**) – while going in the other direction, up and away from **Pd,** we reach games such as Battle (**Ba**) and Chicken (**Ch,** in which two players drive at each other head-on, and must choose to hold their nerve or swerve and be judged "chicken").[14] The various phases of the Cold War then seem something like a dance, a meta-game, played out in moves on this table – with each side often unaware of the other's position, and trying to deceive the other about its own.[15] As long as peace

Ch	BaCh	HrCh	CmCh	DlCh	PdCh	ShCh	CoCh	AsCh	PcCh	HaCh	NcCh
ChBa	Ba	HrBa	CmBa	DlBa	PdBa	ShBa	CoBa	AsBa	PcBa	HaBa	NcBa
ChHr	BaHr	Hr	CmHr	DlHr	PdHr	ShHr	CoHr	AsHr	PcHr	HaHr	NcHr
ChCm	BaCm	HrCm	Cm	DlCm	PdCm	ShCm	CoCm	AsCm	PcCm	HaCm	NcCm
ChDl	BaDl	HrDl	CmDl	Dl	PdDl	ShDl	CoDl	AsDl	PcDl	HaDl	NcDl
ChPd	BaPd	HrPd	CmPd	DlPd	Pd	ShPd	CoPd	AsPd	PcPd	HaPd	NcPd
ChSh	BaSh	HrSh	CmSh	DlSh	PdSh	Sh	CoSh	AsSh	PcSh	HaSh	NcSh
ChCo	BaCo	HrCo	CmCo	DlCo	PdCo	ShCo	Co	AsCo	PcCo	HaCo	NcCo
ChAs	BsAs	HrAs	CmAs	DlAs	PdAs	ShAs	CoAs	As	PcAs	HaAs	NcAs
ChPc	BaPc	HrPc	CmPc	DlPc	PdPc	ShPc	CoPc	AsPc	Pc	HaPc	NcPc
ChHa	BaHa	HrHa	CmHa	DlHa	PdHa	ShHa	CoHa	AsHa	PcHa	Ha	NcHa
ChNc	BaNc	HrNc	CmNc	DlNc	PdNc	ShNc	CoNc	AsNc	PcNc	HaNc	Nc

FIGURE 5.5 Classification of games, based on Bruns, *Names for Games*, in turn based on Goforth and Robinson, *Typology*. The first player has the payoffs of the game of the row label: **Ch**icken, **Ba**ttle, **Hr** (hero), **Cm** (Compromise), **Dl** (Deadlock), **Pd** (Prisoner's Dilemma), **Sh** (Stag Hunt), **As**surance, **Co**ordination, **Pc** (Peace), **Ha**rmony, **Nc** (Concord). The second player has the payoffs of the column label. Adjacent games are related by swapping two adjacent-ranked payoffs. Swapping top-ranked payoffs also "wraps around" the whole table, top to bottom and left to right, while swapping low-ranked payoffs wraps each quarter of the table in the same way.

remains both sides' favourite outcome, we are in Stag Hunt or one of its peaceful mutations in the bottom-right. War is of course possible, but as long as both sides are confident of the other's stability and intentions, it need not occur. A switch by either side – even only in the other's imagination – to preferring to strike first moves the game first to an Asymmetric Dilemma, then to Prisoner's Dilemma, then to the still more dangerous games in the top-left.

In the period of known US nuclear advantage, after 1945, when there was a real US nuclear monopoly, and of effective monopoly between 1949 and the mid-1950s when the Soviets had the bomb but had no obvious means to threaten the continental United States with it, the "game", such as it was, was "Peace" **Pc** or "Concord" **Co** or a mildly asymmetric combination such as **PcCo**. The details don't matter because this is a part of the periodic table in which (*Don't, Don't*) is both Pareto-optimal and the only Nash equilibrium, and *strike* is never dominant. However, American commentators, notably von Neumann, and even the famous British philosopher and later peace campaigner Bertrand Russell, openly advocated preventive war to exploit America's superiority before the window of opportunity closed. In game-theoretic terms, they were arguing that the United States should choose a dominant strategy of *strike*.[16] Fortunately for Moscow, the Truman administration explicitly rebuffed such talk on ethical rather than rational grounds. However, even the mention of such a possibility created a rational source for what the West would call "paranoia" on the part of the USSR. Soviet reluctance to take the good intentions of the United States at face value when the stakes were so high was surely understandable.

In the 1950s, Soviet leaders knew much more about US nuclear capabilities than the United States knew of theirs and were fully aware that they were unable to seriously damage the continental United States in the event of a nuclear war without ICBMs or a functional strategic bomber force of their own, while the obliteration of the USSR in nuclear war became increasingly certain as US strategic forces grew in power. The United States was playing **Sh** priorities, but the Soviets naturally feared that it might desire to use its first-strike capability and thus change its priorities to those of **PdSh**. The Soviets could not participate in a nuclear war except as a victim and their only possible deterrent strategy was to create the opposite impression by deception until their strategic capabilities could develop to provide a credible deterrent.

Thus, in the mid-1950s, Soviet propaganda successfully persuaded the United States that the USSR had the capability to destroy America with nuclear bombers such as the M4 bomber (see Figure 5.6) and intercontinental missiles that it did not in

FIGURE 5.6   The Soviet M4 bomber was intended to present a strategic nuclear threat to the continental United States. Objectively, its limited range and low production numbers made it useless in this respect but wild American overestimation of its capabilities and scale of deployment actually turned it into an effective deterrent weapon in the 1950s. (US Navy)

reality possess. In fact, they had "no deterrent at all".[17] The Soviet space launches beginning with Sputnik 1 in 1957 provided a credible basis for such claims by shattering the faith of the United States in its basic technological superiority and demonstrating that the Soviets must indeed possess an ICBM to lift their space vehicles (though in reality the R7 rockets were impractical as missiles, under the NATO codename SS-6 Sapwood, and built only in the small numbers appropriate for the space programme – see Figure 5.7). The Soviets had achieved basic deterrence by bluff, convincing the Americans that they could not win a nuclear war.

It was an astonishing achievement, but the Soviet success came at a cost in two ways. First, apparently secure deterrence in the short term eliminated the possibility of avoiding a costly arms race. Secondly, the successful deception, resulting in the American perception of imaginary bomber and missile "gaps", diminished Soviet security rather than enhancing it as the United States acted to widen the actual capability gap. In the absence of hard intelligence for the "gaps", and the willingness of the US Air Force to endorse aggressive Soviet propaganda in pursuit of funds and relevance, American public opinion, "inclined to believe the worst", became convinced that the fictional bomber and missile gaps were real.[18] The academics at RAND, alarmed by Air Force intelligence estimates, argued that the Soviets could achieve "first strike" capability soon, destroying America's own nuclear strike capability in a single surprise attack and "winning" the war. RAND theorist Albert Wohlstetter stressed apparent US vulnerability in papers advocating "hardening" US bomber bases and enhanced preparedness,[19] but ultimately the answer appeared to be a massive strengthening of US nuclear forces to the point where any first strike would be shrugged off and US retaliation or "second strike" massive and effective.

In game-theoretic terms, the danger is of perceptions moving from Stag Hunt, relatively benign at least under a belief in one's opponent's rationality and stability, to an Asymmetric Dilemma or a full, symmetric Prisoner's Dilemma, in which instability or paranoia could appear to make striking rational. The Soviets, in their desire to turn an asymmetric dilemma against them (**PdSh**) into a more secure Stag Hunt (**Sh**), were in danger of creating a new and unprecedented reverse asymmetric dilemma from the American perspective, that of **ShPd**, in which it was the Soviets who had the first-strike capability. The actual lack of a Soviet capability became irrelevant to American policy, as US decision-makers were unaware of it. However, at this point in the late 1950s, there was at least the possibility of an emerging stability – each side was playing benign, Stag Hunt priorities, and although there had been, or it had perceived, a first-strike capability in its opponent, its own growth in capability then denied this. Once this is understood to hold, the game becomes Stag Hunt (**Sh**), and for both sides' optimal strategies to be not to strike (the Nash payoff equilibrium) requires merely a moderate belief in one's opponent's rationality, a belief which the stability of **Sh** can itself engender. The Soviets had no intention of initiating a suicidal nuclear attack on the United States, and the Americans, having forsworn first use, assumed themselves to be founding deterrence on the basis of a massive retaliatory second-strike capability. Furthermore, U2 surveillance revealed to the US government some calming evidence of the true situation of Soviet inferiority, at least in terms of strategic bombers. This led to the Eisenhower administration facing damaging public charges of complacency, though possessed of reassuring knowledge it could not share, given the illegal way in which the intelligence was gathered.

The head of Strategic Air Command, General Curtis LeMay, a leading critic of "complacency", even went so far as to characterize the notional Soviet participant in the game as "the gnome in the basement" of the Kremlin,[20] constantly assessing the prospects for a strike on the United States – and likely one day to conclude that conditions were propitious and the time had come. By a peculiar process of inversion, such hawkish Americans wrongly believed themselves to be in the defenceless position occupied in reality by the Soviets. In these circumstances, a dangerous situation could

FIGURE 5.7 (a and b) The R7 rocket, of which this "Soyuz U" was a refinement (a), was a famous and effective space launch vehicle but a poor ICBM when adapted for this role as the SS6 "Sapwood" (b). It existed in small numbers at known sites above ground. It also required 20 hours to be fuelled for action and should have provided no objective deterrent capability. As with the M4 bomber, however, American over-estimation of its capabilities and scale of deployment gave it effective deterrent value in tandem with the bomber. (Wikimedia Commons, (a) photographer unknown, (b) Heriberto Arribas Abato)

(Continued)

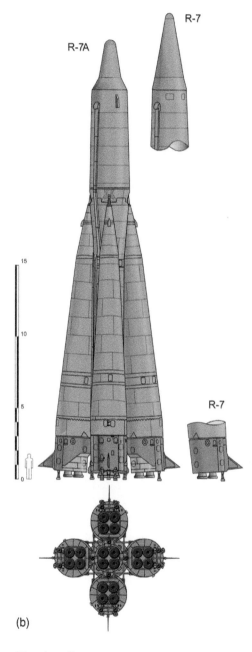

FIGURE 5.7 (*Continued*)

be created purely by accident, as it nearly was on many occasions, or if the American armed forces believed mistakenly that a Soviet strike was imminent or underway. In this eventuality, the United States would assume that a very heavy Soviet attack could be expected and the US war plan, the Single Integrated Operational Plan or "SIOP", simply envisaged the launching of the entire US arsenal, which by the early 1960s included thousands of thermo-nuclear warheads deliverable by ICBM, IRBM, MRBM, SLBM, strategic bombers and nuclear-armed tactical strike aircraft.[21] In this context, the Prisoner's Dilemma game would reach a ter-rible conclusion, with the Soviet role played by the imposing and fictional caricature of US imagination. As game theory demon-strates, these were indeed dangerous times.

PHASE 2: DETERRENT INSTABILITY AND CUBA

The situation of broad but potentially unstable strategic balance ended definitively in the early 1960s and the destabilizing effects made themselves felt almost immediately. Soviet immunity to espionage, which had famously enabled Khrushchev to boast that the USSR was turning out missiles "like sausages", was over-turned by technological progress. By the end of the 1950s, U2 flights had revealed that the bomber gap did not exist except in the sense that SAC was massively preponderant and the Soviets had no credible strategic bomber force. Corona satellite imagery then revealed that the Soviets had a negligible ICBM force as well, possessing in 1960 only four dedicated ICBMs of the type which had launched Sputnik.[22] These were liquid-fuelled rockets which had to be fuelled over a period of hours above ground, and the Americans now knew where they all were. In a RAND briefing, Daniel Ellsberg revealed to an incredulous audience that there was a missile gap, but in America's favour at a ratio of 10:1.[23] The American leadership were now aware that their estimates of Soviet nuclear strength in the late 1950s were wildly exaggerated and that their own nuclear forces greatly outmatched those of the USSR in scale and capability. The key fact was that the Americans now knew that the USSR could not credibly threaten the continental United States with nuclear attack.

The simple possession of this knowledge did not necessarily change the deterrent game and was in some respects potentially beneficial. The United States did not intend to attack the USSR with nuclear weapons, and the knowledge of Soviet weakness greatly reduced the possibility of a massive American nuclear response to an imagined Soviet first strike, which was now known to be impossible in the short term. The Americans also knew that the Soviets would achieve some form of strategic nuclear parity by the mid-1960s but would have no first-strike capability. Thus, the conditions propitious for *détente* – those of Stag Hunt, **Sh** – already existed.

However, at this point Soviet confidence that the Americans were playing **Sh** was still dependent on believing that their deception was working. The Eisenhower administration had kept the intelligence acquired from U2 flights in the 1950s secret, as before Gary Powers was shot down over the USSR they could not reveal this secret source of illegally gathered intelligence and had to silently endure allegations of complacency. They were also unsure before the Corona satellite imagery became available of the extent of the Soviet ICBM arsenal. The Soviets thus remained confident that their previously successful bluffing still exerted a real effect on US calculations. However, in the 1961 dispute over Berlin the Kennedy administration revealed publicly that they knew the truth about the nearly non-existent Soviet ICBM force, in a move which was interpreted by Moscow as a threat of strategic nuclear first strike if it moved against West Berlin. This revelation by the Americans that they knew the asymmetric dilemma **PdSh** favoured them changed everything fundamentally from the Soviet perspective.

By the early 1960s, the USSR had built up a sizeable and modern MRBM and IRBM force which was a major threat to European NATO, but none of these weapons could reach the United States and the USSR was now vulnerable to strategic nuclear blackmail by the revealed threat of an effective American strategic nuclear monopoly. Nor did the Soviets assume that the United States would not use the opportunity to attack them, given their previous experience of such a situation and the known views and character of US military leaders such as LeMay. They would achieve strategic deterrence, and thus a reversion to true **Sh**, in a matter of years, but until then faced a clear window of vulnerability, in which the Americans were confident that their homeland was effectively immune from nuclear attack. There was, however, the possibility of closing the window by placing MRBM and IRBM weapons in Cuba, which would reward previous investment in these weapons and instantly provide the deterrence which would otherwise take years to acquire. This was only one reason, but certainly a compelling one, for the installation of such warheads in 1962.

Unfortunately for Khrushchev, a move which promised to restore strategic stability in fact did the opposite. In a literal sense, it worked as a concept. The USSR would gain the capacity to strike the continental United States. The problem was that the balance tilted rather in its favour, as the flight times of Cuba-based missiles to US targets was very short, creating the possibility of a decapitation strike. Had the missiles become operational before they were discovered, the asymmetric dilemma would have been reversed, favouring the USSR in a genuine **ShPd** which had previously only been a figment of US over-imagination in the all-too-effective Soviet bluff. To vitiate in a single act the Stag Hunt equilibrium would certainly be seen by the United States as an act of war, which was a powerful reason not to deploy the missiles in the first place. But discovery before activation created an acute problem, for the Americans could not allow the latter to happen.

In the resonant American mythology of the incident, US Secretary of State Dean Rusk reported that he had said "We're eyeball to eyeball, and I think the other fellow just blinked", a conscious claim that the game being played was a gaze-holding "Chicken" – in which the only Nash equilibria are indeed for one or the other player to "blink". However, when one considers the

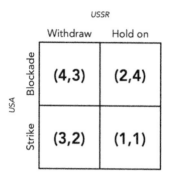

FIGURE 5.8   Cuba (NcCh).

options open in the short term – for the United States, air strikes (before a possible invasion of Cuba) or a continued blockade without striking; for the USSR, to hold on or withdraw – the game is something like Figure 5.8.

The United States' most favoured outcome is Soviet withdrawal without air strikes, its second favourite Soviet withdrawal following air strikes – the crucial ordering is that to strike and cause withdrawal is better than to allow the missiles to become operational. If the Soviets hold on, one could argue that the game is inadequate and will be superseded, so the US 1 and 2 options could be either way round – we have assumed that in this case the United States would rather not have shot its bolt. For the Soviets, the favourite outcome is to hold on without the Americans striking, followed by withdrawal without air strikes. If the USA strikes, the USSR prefers to withdraw.

This game is a variation of Chicken (**NcCh**) in which the only Nash equilibrium is (*Blockade, Hold on*). However, it is not stable to a reversal of the Americans' two least favoured options. If the Russians hold on and the United States prefers to strike, the (*Blockade, Hold on*) equilibrium disappears and there is none to replace it – we have no natural optimum, and we are in the realm of Brams' "theory of moves" (**ShCh**).[24] If the Russians also exchange their least favoured options, and prefer to hold on even at the risk of air strikes, then we arrive at the **ShPd** form of asymmetric dilemma and (*Strike, Hold on*) becomes an equilibrium. Brams even prefers a variant in which the Americans prefer (*Strike, Hold on*) over (*Blockade, Withdraw*), for reasons of honour.[25] If the reader is now muddled, they may assume the politicians were in no better state. The game is uncertain, and no sane person would stake the world on the correctness of their understanding of it.

Khrushchev's initiative was thus a reckless gamble with deterrence, in stark contrast to previous Soviet caution, aimed at completely overturning the existing game and therefore dangerously destabilising. Successful deployment of the missiles would have left the United States with no valid strike option and very limited deterrent capability, if the Soviet missiles could not be destroyed before they were launched. In the midst of the general atmosphere of shock and disbelief when the missiles were revealed, US secretary of Defence McNamara made the calming point that deployment only brought forward the date of Soviet strategic parity, thus creating no real change, and Ellsberg argued that decapitation was not a useful option for the Soviets as delegation of authority meant that the full SIOP response would happen anyway in this event.[26] This overlooked two factors. First, the Soviets had no idea that a decapitation strike would not paralyse US decision-making. Both superpowers assumed that the political leadership of their respective countries was in full control of nuclear release – although both were wrong, as the crisis revealed. Secondly, at no point in the Cold War was the political leadership of either superpower willing to accept even the possibility of a surprise attack designed to kill them. They equated their own survival with that of the nation.

When the weapons were discovered before activation, the Soviet initiative backfired. The United States could not and would not allow the missiles to be activated. Air strikes might have been a conventional attack in the first instance but there were no illusions on either side about the possibilities of nuclear escalation even without knowledge of the latitude available to each other's junior commanders. Khrushchev had created the situation the Soviets had sought to avoid throughout the 1950s by inviting an irresistible American attack. Given the uncertainty in the nature of the game, the best American policy was to convince the Soviets immediately that they would strike rather than let the missiles be activated, and strike whatever the Soviets did – whether this were true or not. The United States was willing – or had to appear to be willing – to accept the risk of escalation, and Kennedy announced on television that any use of US nuclear weapons would be the "full retaliatory response" involving the complete SIOP attack on the USSR and its allies.

The Soviet leadership knew that without the weapons on Cuba, it could not effectively respond to American attack with the handful of strategic weapons it possessed in the USSR. They also undertook no measures of nuclear mobilization, as in addition to the dangers of escalation they felt the United States to be prepared for pre-emption if they believed they were about to be attacked. Unable to risk this outcome, Soviet strategic rocket forces were effectively paralysed. The Americans had thus, in effect, called the Soviet bluff. As an application of game theory, it is hard to over-state Khrushchev's blunder. He had only to wait for a few years until Soviet ICBM deployment realised MAD and turned the bluff of the 1950s into reality, creating the genuine payoff equilib-rium of peace and cooperation in the Stag Hunt, but had decided that the risk of waiting outweighed the risks of action. Instead, in circumstances in which Soviet strategic weakness had become known, he pushed the United States into a time-limited decision-making process in which a conventional air strike was inevitable, and a full first strike on the Soviet Union was at least a likely esca-lation and possibly a rational and even attractive option. Removal of the missiles, and quickly, became his only option.

The missile crisis was chastening for all involved, and although it ushered in greater contact between the superpowers and some measures of nuclear restraint, the eventual achievement of strate-gic parity for the USSR with weapons based on its own territory finally achieved the desired Nash payoff equilibrium with a valid MAD doctrine. It was acknowledged that the missile crisis occurred at the very end of the US period of nuclear dominance, for the first generation of effective Soviet ICBMs was in deployment in October 1962, but not yet in the numbers required to inflict terminal damage on the continental United States.

Concerning features of decision-making before MAD could, however, reappear if MAD were to break down, as it did in the mid-1970s. The game of the 1950s had proven to be viable in terms of deterrence because Soviet success in simulating MAD provided a stable equilibrium. A game in which both sides had an illusory fear of first-strike vulnerability would be much more dangerous, however, and this situation was created in the renewed Cold War of the late 1970s and early 1980s.

PHASE 3: THE END OF DÉTENTE AND THE ROAD TO ABLE ARCHER

The Able Archer crisis replicated many of the features of the earlier crisis but was in important respects even more dangerous. Once again it placed great pressure on the nuclear decision-making processes of the superpowers. In the decade leading up to 1962, weakness in American intelligence gathering placed the United States in a counterfactual mindset which did not reflect objective reality. In terms of the adversarial stance with the USSR, game theorists at RAND assumed themselves to be in a simple two-player game but failed to appreciate the crucial distinction between notional and real Soviet participants. Fortunately, this situation was sustainable because the Soviets were conscious of their true weakness and so declined to participate as the powerful aggressor of American imagining until their true weakness was revealed and Khrushchev overplayed his hand in Cuba attempting to level the playing field.

The era of *détente* was, in game-theoretic terms, a re-setting to Stag Hunt. It reflected an acceptance by both sides of the concept of Mutually Assured Destruction (MAD) after the Soviet achievement of parity in strategic nuclear weapons with the United States. It was clear by the second half of the 1960s that while each of the superpowers could obliterate the other with a nuclear strike, neither could hope to escape similar devastation in the inevitable counterstrike, now delivered by a "triad" of ICBM, aerial and submarine launched weapons. Given that the existing arsenals of the superpowers were sufficient to deliver Armageddon many times over, the advantages in removing this costly "redundancy" with nuclear arms limitation and reduction treaties were obvious to both sides. *Détente* involved a conscious move towards closer interaction between the superpowers and thus a more managed Cold War environment reflected in such outcomes as the SALT treaties, the ABM Treaty and the Helsinki Accords in the early and mid-1970s. These initiatives appeared to defuse the arms race and create an environment in which the proliferation of nuclear weapons could be halted and then possibly reversed. Equally importantly the possibility of a nuclear conflict arriving by surprise, by accident or through the escalation of an international crisis was seen to be reduced.

In terms of game theory, this was a truly stable situation. Here, both sides understood each other's position and only one course of action was rational for each side, namely cooperation at the strategic nuclear level, making the payoff equilibrium stable. This was a strictly functional relationship, however. It did not switch off the Cold War but ensured that while *détente* lasted it would be fought at a less dangerous level. Minimising the risk of nuclear catastrophe did not mean abandoning Cold War objectives and the dominance of Cold War ideologies. In the Soviet view, the "correlation of world forces" would continue to move in the direction of world communism more effectively once nuclear threats were removed, while the United States assumed that its model of liberal democracy shaped the natural direction of historical travel. Thus, conflict continued. American gains in the Middle East as a result of the Yom Kippur War of 1973 were resented in Moscow, but perhaps more significantly a string of Cold War reverses for the

West, including the first oil shock, defeat in the Vietnam War and communist advances in Africa, convinced many that the West was losing the Cold War.

The Able Archer incident occurred during the so-called Second Cold War, a term which reflected a new and dangerous period in international relations following the progressive unravelling of *détente* from the mid-1970s. The introduction of a new generation of Soviet IRBMs in Europe, the infamous SS-20s, and a simultaneous build-up of Soviet conventional forces on NATO's Central Front were matters of particular concern to politicians both on the right and, increasingly, in the centre.[27] They considered Soviet behaviour menacing and came to perceive international agreements with Moscow in the context of appeasement, employing renewed rhetoric of rearmament and counterattack against communism, or from the Soviet perspective mounting an effort to "combat the laws of historical development".[28]

In this climate, the Soviet incursion into Afghanistan in 1979 was perceived by the US administration as an ominous event, and the previously dovish President Carter and other US politicians reacted forcefully. The US Congress refused to ratify the newly signed SALT II nuclear arms reduction treaty[29] and Carter, fearing a breakdown of deterrence through Soviet attainment of a first-strike capability, sought from his advisors a more robust plan for the contingency of nuclear war.[30] The new policy, encapsulated in Presidential Directive 59 [PD 59], sought to enhance the deterrent effect of US policy by targeting in principle that which Soviet leaders most valued: their own lives and power. Thus, the United States began to think in terms of a "decapitation" strike aimed at the Soviet leadership in their bunkers.[31] This perception was accurate in terms of Soviet fears, but has been described as the result of a "mechanistic theory of deterrence"[32] driving the Americans, and it destabilised the nuclear balance and produced the opposite of a deterrent effect.

The Soviet leaders did not consider themselves to possess or even to approach possession of a first-strike capability. This being so, the creation among them of a realistic fear that they might be subject to it themselves raised multiple dangers to peace. Critically, they were no more willing to accept the possibility of decapitation in the late 1970s than the Americans had been in 1962. The disintegration of *détente* which accompanied and in part created PD 59 indicated a new instability. In game-theoretic terms, this was the instability of **Sh**, in which the belief that the other possesses first-strike capability and the fear that they might use it combine to destabilize the payoff equilibrium of peace, moving the game to the asymmetric and then full Prisoner's Dilemma **Pd**.

This new situation echoed the 1950s, certainly in terms of US misunderstanding of Soviet intentions and gross overestimation of Soviet capabilities. However, in the emerging late 1970s reality neither side understood the position of the other, a fundamentally new and dangerous situation. As before there was, from the US perspective, the notional and highly aggressive Soviet superman the United States believed itself to be facing, in contrast to the real and fearful Kremlin. Now, however, the Soviets were also deluded as to American intentions and imagined themselves to be facing not only a mighty foe but also an openly aggressive opponent, not only creating the opportunity for a decisive, decapitating first strike but apparently planning to do so.

This mirrored the US error of the 1950s and did not reflect the true situation of the Americans seeking to reinforce deterrence by creating a credible threat to the Soviet leadership. As with Cuba, however, creating a plausible fear of a decapitation strike proved to be the single most dangerous way to destabilise the nuclear equilibrium, as it undermined MAD and created the possibility of rational pre-emption. Given the precedent, it is difficult see how US planners of the 1970s considered that frightening the Soviet leadership in this way could possibly be a good idea. In this situation, neither side envisaged itself initiating a nuclear attack as an act of policy, but as before the real danger lay in accident or misconception of events which might leave one side feeling extreme pressure to launch strategic weapons pre-emptively, which, unlike in the 1950s, was now a viable option for both sides. This had long been recognised as a likely trigger for war.

The Able Archer crisis was both short and long – brief in duration, but the culmination of decades of flawed planning. The imminent arrival of Cruise and Pershing II "Euromissiles" (see Figures 5.9 and 5.10) in European NATO in 1983 provided the proximate cause of the crisis along with a series of random events

FIGURE 5.9 The Pershing II IRBM was a new weapon due to be deployed at the end of 1983 as one of the two "Euromissiles" designed to counter Soviet deployment in the 1970s of the SS20 IRBM in eastern Europe. Understood by the Soviet leadership to be capable of destroying their command bunkers while they were inside them with minimal warning, it created a fear of "decapitation" similar to that felt by US leaders in the Cuban Missile Crisis and was similarly destabilising in the deterrence context. (US National Archives ID 6385395)

FIGURE 5.10 The Air launched Cruise Missile (ALCM) was the second and lesser of the two new "Euromissiles" intended to counter the deployment of Soviet SS20s in Europe. Effectively a "stealthy" descendent of the German V1 its subsonic speed made it less alarming than the Pershing II to Soviet leaders and it did not destabilise deterrence equations. (USAF National Air and Space Museum)

which appeared to the Soviets to be connected and indicative of imminent attack. Unlike Cuba when the US response had been immediate and vocal, the Soviets did not communicate their immediate fears through any channel, leaving the Americans unaware of them and instead projecting onto the Soviets the traditional image of strength. The absence of dialogue magnified the effects of intelligence failure on both sides.

The basis of Soviet calculations was a mix of good and poor intelligence work. The Soviets obtained a copy of SIOP 6 through espionage and were as alarmed as the strategy intended them to be, in fact more so. Their intelligence revealed that the Americans knew where the bunkers of the leadership, "time-urgent hardened targets",[33] were and could probably already destroy them, as they also knew that US strategic weapons generally possessed greater accuracy than their own. They also knew, and the Americans did not, that their ability to detect a pre-emptive

US strike was very limited and would remain so until the new "Perimeter" early-warning system came into service in the mid-1980s. The Pershing II could reach Moscow from West Germany in 10 minutes – "roughly how long it takes some of the Kremlin leaders to get out of their chairs, let alone to their shelters."[34] The Soviets were properly impressed by the imminent capability of Pershing II to hit deep bunkers with accuracy and destroy them, but they were also aware that they were in immediate danger. The US planners thus worked from a mistaken premise. They assumed SIOP 6 would deter attack if the Soviet leadership knew they could not survive a general nuclear war. However, this would only be true if the Soviets believed that the Americans did not intend to attack them – a belief that declined steadily through the early 1980s. In January 1983, newly appointed General Secretary of the Communist Party and Soviet Premier Yuri Andropov told Warsaw Pact leaders that it was "difficult to distinguish between what constitutes blackmail and what constitutes a genuine readiness to take the fateful step".[35] The Americans developed a credible capability to launch a first strike that they did not intend to use, while the Soviet leadership accepted the truth of American capabilities but came to believe that they *did* intend to strike at a propitious moment. In these circumstances, the only possible Soviet responses were to launch as soon as an attack was detected, "launch on warning", or to launch pre-emptively if intelligence revealed that an attack was certain and imminent.

In the absence of effective early warning, the Soviets developed an automated intelligence gathering system known as Operation RYaN (a Russian acronym for "nuclear missile attack"). This was devised by Andropov, then head of the KGB. The RYaN system would integrate multiple sources of intelligence which might provide any evidence of an imminent NATO attack, famously down to the time the lights were still on in Western defence ministries on Friday evenings, into a database providing a computerized analysis of the level of NATO nuclear mobilization. This would compensate to some extent for the lack of a technological early warning system and provide some indication of an imminent Western attack, reducing the extent of surprise NATO could achieve.

The significance of RYaN has become the subject of controversy, in the sense that historians and historical actors disagree about the extent to which it was a significant component of Soviet intelligence gathering or heightened the threat of nuclear war. Those minded to highlight the dangers inherent in the system stress its evident shortcomings for intelligence gathering. It relied on faulty assumptions of Western behaviour. It allowed agents no latitude to interpret the data they were asked to collect and was prone to confirmation bias in two ways. The knowledge that leaders were looking for signs of Western aggression led agents to provide it for careerist reasons, and in addition the belief of Andropov and his colleagues in fundamental US hostility and imminent NATO aggression predisposed them to interpret data pessimistically.

Others, however, argue that the importance of RYaN has been grossly overstated. It has been alleged that it was "a research and development project rather than a serious source of insight into NATO's thinking"[36] which was not considered a functional part of the Soviet early warning system and "would not have been trusted by the intelligence agencies or the policy makers of the Warsaw Pact to make the decision to launch a pre-emptive nuclear strike on the west."[37] Other critical commentary has come from allied Warsaw Pact intelligence agencies and within the Soviet Union from the military intelligence organisation GRU, separate from the KGB and indeed a rival to it,[38] which naturally stressed the importance of purely military intelligence.[39] Former Soviet agents and military personnel have also downplayed in reminiscence the idea that faulty intelligence gathering precipitated a war scare in 1983,[40] though of course to acknowledge that it did would reflect badly on them. There was also a tendency, from those not involved in it, to pour scorn of the idea of a functional computerised intelligence gathering system.[41]

Ultimately the degree to which the possibility of a war scare based on faulty intelligence and policy formulation in 1983 is taken seriously tends to divide along Cold War lines, with Western actors of the time taking the idea seriously and Warsaw Pact counterparts downplaying it. However, it is clear that the fundamental weakness of deterrence under perceived first-strike threat and misplaced fears based on lack of intelligence data constituted considerable dangers in themselves, reinforced by the breakdown of meaningful contact between the two blocs. In fact, no one person in the world possessed the informed perspective required to perceive the danger, except possibly the double agent Oleg Gordievsky, a KGB operative and British double agent working in the Soviet Embassy in London, who was aware of intensifying intelligence efforts under RYaN. However embryonic the system might have been, RYaN was certainly operational by 1983 and its alert status was raised in February and again in June, ordering intensification of intelligence efforts by Soviet agents aimed at detecting "any sign of a coming surprise attack."[42] Gordievsky made repeated and initially unsuccessful efforts to warn his British handlers of the level of alarm present in the Soviet leadership and the dangers presented by "a potentially lethal mix of Reaganite rhetoric and Soviet paranoia".[43] The British foreign secretary Geoffrey Howe observed that Gordievsky's intelligence demonstrated to him "that the Soviet leadership 'really did believe their own propaganda'".[44] Equally disturbingly, President Reagan later expressed surprise, even shock, that they believed his.[45]

The imminent deployment of Euromissiles had been the major Cold War issue of the period with extensive peace movements in Western Europe, particularly in the United Kingdom and West Germany, seeking to prevent their deployment assisted by an intensive Soviet propaganda campaign. These efforts proved fruitless, with secure right-wing governments in both countries determined to deploy the weapons on schedule in the autumn of 1983. To this setback was added the Strategic Defense Initiative programme (SDI, sometimes referred to as "Star Wars") in 1983, which to the Soviets seemed to threaten not only the stability of MAD with a successful US missile defence system but also the possibility of offensive space-based beam weapons capable of instant and unstoppable attacks on the USSR. To the Soviets, it also seemed that the hostility of the United States was increasing, perhaps to a planned schedule. President Reagan's "Star Wars" speech of 23 March 1983 had been preceded by another in the same popular cultural context on 8 March to a religious audience in which the President denounced the USSR in unprecedented terms as the "evil empire" and "the focus of evil in the modern world". The venomous American reaction to the shooting down of Korean Airliner KAL007 on 1 September 1983, which had strayed into Soviet airspace and been misidentified (it was later established) as a USAF RC135 reconnaissance aircraft, was interpreted as corroborating the "Evil Empire" speech with the assertion that Soviets "marched to a different drum". The increasingly paranoid Soviet leadership was then faced with the US invasion of Grenada, accompanied by much signals traffic between the United States and United Kingdom. The accelerating tempo of events and incorrect interpretation of RYaN data led the Soviets to conclude that forthcoming Able Archer command post exercise would make an excellent cover, "maskirovka", for the initiation of a nuclear attack, whether or not they believed it would happen. Though the exercise was held annually, new features in terms of encrypted command and control and the proposed involvement of heads of government raised Soviet suspicions. In terms of game theory, all elements were present for a Soviet pre-emptive strike or a launch on warning except the trigger.

THE NUCLEAR STAND-OFF GAME

This situation is more complex than the traditional, static two-person games of RAND.[46] To model, it we create a parametrized variant of Stag Hunt (Figure 5.11) and allow varying payoffs, to

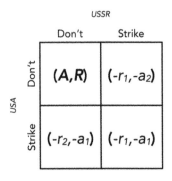

FIGURE 5.11   The Nuclear Stand-Off.

take account of changing circumstances. We assume that the values of a peaceable economy A to NATO ("America") and R to the Warsaw Pact ("Russia") are always large and positive. *Strike* means a first strike, while *don't* may mean just that (if the other side does the same) or a second strike (if the other side has struck first). In a nuclear war, we assume that each side causes damage in proportion to its nuclear arsenal, similar to the Lanchester model – but that a first strike causes different, and potentially much greater, damage than a second strike (a_1 compared to a_2 for a strike by NATO, r_1 compared to r_2 for the Warsaw Pact).[47] It's clear that by far the best thing all round is no-strike on both sides, just as it should be. To be mathematically precise, there is no mixed strategy in which each side optimizes its payoffs by choosing to strike with non-zero probability p_R (for Russia) or p_A (for America), again as we would wish. Further, the ordering of outcomes is no longer strict, for two outcomes are ranked equally for each side: if the other side has a strategy of striking first, we do not assume any advantage to our own side of doing so.[48] This is in the spirit of a firm and fatalistic belief in the assuredness of mutual destruction in nuclear war, but it does not rule out pre-emption – which we shall introduce in another way. Peace, (*Don't, Don't*), is Pareto-optimal and a Nash payoff equilibrium, as we expect, while (*Strike, Strike*) is no longer a strict equilibrium. There are no dominant strategies.

Could it ever be rational for Russia to launch a first strike? As noted above, if the game is analysed conventionally, the answer is no, not even as a possibility. But suppose that Russia irrationally thought that the probability of an imminent American strike was p_A.[49] Should they get their retaliation in first? In the game as described, the answer is "no" – their payoff is never improved by striking first. But standard two-person games of nuclear standoff have a flaw: both sides cannot make a first strike.[50] If the Russians strike then the Americans can only make a second strike (and *vice versa*), and the Russian payoff is $-a_2$. If the Russians do not strike, then their payoff is dependent on whether the Americans strike or not and is

$$-a_1 p_A + (1 - p_A)R.$$

The two are equal at the critical value

$$p_A^C = \frac{R + a_2}{R + a_1}$$

(which is always positive and less than one). Thus, if Russia ever believes that an American strike is imminent – that NATO will strike immediately with probability p_A higher than the critical probability p_A^C – then, in contrast to all rational analysis based on the standard two-person game, it is rational for Russia to strike first. We can think of the developing crisis as a tension between Soviet assessment of NATO strike probability p_A, as deduced from RYaN, and the critical threshold p_A^C given by the formula above.

Now let us wind back and look at the development of this game though the decades. Suppose the Soviets begin in good times. We might expect that p_A would be low, much less than one, while p_A^C would be high, close to one, because of the strength of Soviet society and economy R, the size of the NATO nuclear arsenal a_1, and the low capacity $a_1 - a_2$ of a Soviet first strike to mitigate a US strike. With p_A close to zero and p_A^C close to one, it is never rational for Russia to strike.

The game is symmetric, and exactly the same argument can be developed for the United States – they should launch a first strike if their estimate of the Russian first-strike probability

$$p_R > p_R^C = \frac{A + r_2}{A + r_1}.$$

If behaving rationally, and when A is large, the Americans are highly unlikely to estimate a value for p_R^C that is much less than one.

But things change. One might expect the Russian economy R – a measure of economy, society and politics – to show slow, secular (long-term), steady decline – perhaps accelerating, if economic bad news builds a sense of impending economic crisis and this feeds back into the value that society and its political leaders place on peace. Without the anchoring effect of large R, keeping the value of p_A^C close to one is now much more dependent on a_1 and a_2 remaining about the same.

Now consider the deployment of Pershing II and SS20. Both deployments increase the sizes of the nuclear arsenal – a_1 and a_2 for America, r_1 and r_2 for Russia – which at first sight appears to raise the critical thresholds, reinforcing MAD and making nuclear war less likely. This may indeed be true of SS20, which does not directly threaten American cities, although by threatening Europe it reduces the value to NATO overall of a second strike relative to a first. But Pershing II, with its capacity in Russian perception for a decapitating first strike, increases a_1 much more than a_2 and thereby reduces p_A^C.

As long as the Russians do not seriously believe America may strike, this reduced value of the threshold p_A^C is not dangerous, for p_A is small, and nuclear war only happens when $p_A > p_A^C$. But the behaviour of p_A is very different from that of p_A^C. Remember that it is an estimate of the imagined likelihood of an American strike – imagined by possibly paranoid people in committee rooms and command bunkers, and changing in days, hours, minutes or even seconds. It is likely to change noisily, apparently randomly, with sudden small or large jumps. All it has to do for nuclear war is to exceed the threshold p_A^C – and once is enough.

Consider the position of the Russian leadership. They must – even if only implicitly – estimate a_1 and a_2. They come up with an implied p_A^C which is high but which, as noted above, has been newly reduced by their leadership's concern for their own lives. Then they begin to use RYaN and other information to construct, implicitly, p_A – although p_A may be different in the individual minds of politburo members and the collective minds of the KGB and GRU. The combined, effective p_A may well lurch around randomly. But Russian ideas about the Americans' view of them, about the Americans' p_R and p_R^C, also feed into the Russian estimate of p_A. About A, they perhaps know or care little. But concerning the American capacity to diminish the Russians' own second-strike capability, there is, in the early 1980s, a perfect storm. Intermediate-range missiles raise r_1 more than r_2, at least in European perceptions. More hawkish Americans may well believe that the difference is increased still further by SDI – and this may be an even more alarmed and convinced belief on the Russian side, where SDI's efficacy is unknown. So, in a febrile atmosphere of misunderstanding, p_A is vastly overestimated compared to its true, tiny value, and its random behaviour could easily combine with the recent downward step in p_A^C to create a moment when $p_A > p_A^C$ and a Soviet first strike becomes a real possibility.

The variables of the game are rather ill-defined in terms of real-world measurables, but we can nevertheless attempt to fit the game to data. Figure 5.12 shows what happens if one uses the total number of warheads for a_1 and r_1, and half of this for a_2 and r_2, and GDP for A and R.[51]

This doesn't incorporate a first-strike advantage, yet it still broadly mirrors our understanding of the developing situation. If we construct an approximate timeline for first-strike advantage, the dangers accompanying the introduction of SS20 and Pershing II are made more stark (Figure 5.13).[52]

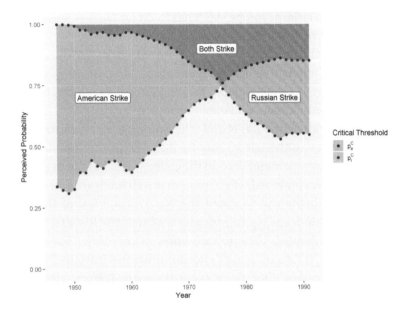

FIGURE 5.12 Perceptions of opponents' first-strike probabilities. If US perception of the probability of a Soviet first strike rises above a threshold value (red dots), it becomes "rational" (within the game) for the United States to strike first. Similarly, the USSR should strike first if it believes the probability of a NATO first strike is above the blue dots. Thus, the blue and red regions give measures of the danger of NATO or Soviet first strikes. (For Figures 5.12–5.14, economic data is real GDP per capita calculated against a 2011 benchmark and is drawn from the Maddison Project Database, version 2018 [Bolt *et al.* 2018]. Nuclear arsenal data is from Our World in Data [Roser *et al.* 2013].)

We can relate the overall risk of nuclear war in our game to the notion of the "Doomsday clock". The risks of a nuclear first strike are that either p_A or p_R – perhaps lurching around somewhat randomly, like a stock-market price – stumble into the gaps between the critical values and one, to the sizes of which the probabilities of a first strike are then proportional, so that the chance of avoiding nuclear war is something like $p_A^C \times p_R^C$. As long as p_A^C and p_R^C remain close to one, it is vanishingly unlikely to be rational

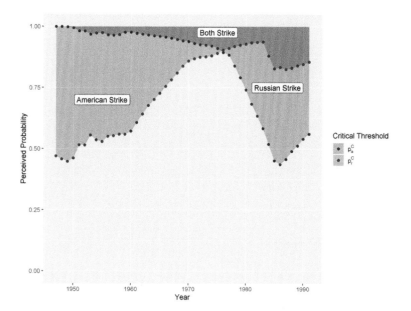

FIGURE 5.13 Perceptions of opponents' first-strike probabilities, with time-varying first-strike advantage.

to initiate a nuclear strike, the hands on the clock stay well away from midnight, and we have the effective peacekeeping mechanism of MAD. Figure 5.14 takes $p_A^C \times p_R^C$ from Figure 5.13 and plots it alongside the official Doomsday clock time-to-midnight. The movements of the two plots are broadly similar – and the grave dangers of the early 1980s in the model are both anticipated and exaggerated by the clock.[53]

The Soviets were mentally prepared for war and accepted its likelihood. All that was required was compelling evidence for pre-emption. As the United States was not about to attack, this should have ended the game, but the possibility of accident or misinterpretation remained as real as it always had been, and the initiation of the Able Archer exercise may have produced the most propitious conditions for pre-emption that the Cold War had ever seen as the procession of events pushed the Soviets towards

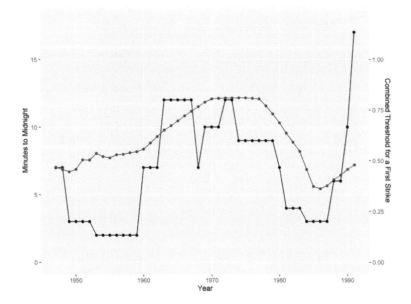

FIGURE 5.14 Doomsday clock and a measure $p_A^C \times p_R^C$ of the probability of nuclear war.

launch. Warnings from Soviet embassies triggered the move of Soviet forces to alert status (though a mistaken return KGB message announcing matching NATO mobilisation was fortunately not transmitted up the chain of command).[54] The measures taken, particularly to move and arm tactical nuclear attack aircraft, were visible to NATO. This finally alerted the Americans and brought them into the game. At this point, deterrence had essentially failed – indeed, in the extreme case, when p_R^C and p_A^C hit zero, the nuclear stand-off becomes, in its preferred outcomes, Prisoner's Dilemma. The situation had come down to the actions of individuals, particularly air force Lieutenant-General Perroots, who found himself in a real and possibly terminal game of Prisoner's Dilemma with the Soviet Union. But Perroots declined the invitation to move away from (*Don't, Don't*), and his decision not to match the Soviet increase in alert status was the definitive act which prevented further escalation.

The Able Archer war scare was in a real sense the culmination of the nuclear Cold War. Afterwards, tempers cooled, and when Gordievsky's warnings of Soviet fears were believed a change in attitude was underway. President Reagan had become pessimistic about the future and made a mental adjustment in favour of arms control. The game returned to the nuclear stand-off, and fictive beliefs in the other as reflexive aggressor (that is, in $p_A > 0$ or $p_R > 0$), resulting in an invisible and unprecedented risk of accidental disaster, were returned to zero by quite modest doses of realism. Equilibrium had been restored to the model and it was again clear that only way to reduce tension permanently was to enact measures of arms control – as was achieved with astonishing speed in the Intermediate-Range Nuclear Forces (INF) treaty of 1987–1988.

The historian would like to have a clear answer to the chapter's title question, and the reader will have noted that we do not attempt to quantify one. We provided a simple model, within which we could have attempted to quantify a RYaN-informed p_A against strike threshold p_A^C, but no sane person would stake the world on it. Instead, we return to our point about the streetlight effect – that there is a danger of using models because they are available rather than because they are right or enable a better decision to be made. We would certainly argue that game theory at least did not help to avert nuclear war.[55] Rather, in the end, the lesson is a humane one – that mutual awareness and understanding have to be worked for but are always beneficial. Even when deception, guile and bluff are real possibilities, it helps to talk, and to think oneself into one's adversary's position. The long road to trust – the only sure way out of a dangerous game – can only be travelled through doing so.

NOTES

1. This, and the shorthand for it provided by "co-orientation theory", forms the basis of the analysis of Able Archer in Burriss, "Slouching toward nuclear war".
2. Poundstone, *Prisoner's Dilemma*.

3. Adamsky, 1983 Nuclear Crisis, p6.
4. This imagines a drunken man who has lost his keys. A passer-by observes the man searching for them under a streetlight, and offers to help, asking "where did you lose them?" "Over there", the man replies, pointing to a dark alleyway. "So why are you looking here?" "Well, I can see here; I can't see over there". This is an example of a broader class of "availability biases". For its effect in medicine, see Freedman, "The Streetlight Effect".
5. The real events form a backdrop to the fictionalized television series "Deutschland "83".
6. Scott, L., "Intelligence and the Risk of Nuclear War", p760.
7. Wells, "The Bomber Gap", p965.
8. Poundstone, *Prisoner's Dilemma*, p.8.
9. One can arrive at these numbers by adding four to the (negative) years of the jail terms.
10. Poundstone, *Prisoner's Dilemma*, p39.
11. There exist "games with incomplete information" or "Bayesian games" in which, for example, one player might think that the other is, with different estimated probabilities, playing one of a number of different games. Typically, however, to analyse these, one needs to know each player's payoffs in each possible game, not merely their rankings.
12. Barrass, "Able Archer 1983", p. 25.
13. For the mathematical-minded, 78 is the number of independent entries in a symmetric 12×12 matrix and is the number of distinct games if we don't distinguish which player is which. The classification is due to Goforth and Robinson, who used numerical labels for the games. We use the less mathematical but more humane labels of Bruns, *Names for Games*.
14. Notice, however, that the matrix "wraps around", both as a whole (by swapping 3 and 4) and in each quadrant (by swapping 1 and 2), as mentioned in the table's description. This gets pretty tricky, so we refer the interested reader to the original text.
15. Bruns, "Escaping Prisoner's Dilemma", is a treatment of game mutations and moves on the periodic table of games, very much in this spirit.
16. In Bruns' *Names for Games*, this is **DIAs**, or something close to it.
17. Ellsberg, The *Doomsday Machine*, p163.
18. Preble, "Who ever believed in the Missile Gap?", p802.
19. Kaplan, *Wizards of Armageddon*, pp118–121.
20. Kaplan, *Wizards of Armageddon*, pp142–143.

21. Ellsberg, *The Doomsday Machine*, pp97–99.
22. ICBM is "Inter-Continental Ballistic Missile"; IRBM and MRBM refer to missiles of medium (less than 2,000 miles) and intermediate (2,000–3,500 miles) range.
23. Ellsberg, *Doomsday Machine*, p164.
24. Brams, *Theory of Moves*. Now there is no equilibrium of any kind, and the two sides' preferences cycle endlessly around the payoff matrix.
25. Brams, *Superpower Games*.
26. Ellsberg, *Doomsday Machine*, p187.
27. Manchanda, "When Truth is Stranger than Fiction", p113.
28. Statement by Yuri Andropov to the Warsaw Pact Political Consultative Committee, Prague, Czechoslovakia, January 4, 1983, in *The Able Archer 83 Sourcebook*.
29. The United States did, however, abide by its provisions.
30. A destabilising situation in terms of consequences for deterrence is discussed in, for example, Carlson and Dacy, "Sequential Analysis of Deterrence Games".
31. Adamsky, "The 1983 Nuclear Crisis", p11.
32. Mastny, "How Able was Able Archer?", p110.
33. Scott, L., "Intelligence and the Risk of Nuclear War", p. 762.
34. Herbert E. Mayer, Chairman of the National Intelligence Council, quoted in Jones, "Countdown to Declassification", p54.
35. Statement by Yuri Andropov cited above.
36. Miles, "The War Scare that Wasn't", pp103–104.
37. Miles, "The War Scare that Wasn't", p107.
38. Barrass, "Able Archer 1983", p8.
39. Manchanda, "When Truth is Stranger than Fiction", p119.
40. See particularly Miles, "The War Scare that Wasn't" for an emphatic attempt to debunk Able Archer as a war threat based on interpretation of Soviet and Warsaw Pact sources.
41. Miles, "The War Scare that Wasn't" pp106–107.
42. Manchanda, "When Truth is Stranger than Fiction", p119.
43. Andrew and Gordievsky, *Comrade Kryuchkov's Instructions*, p86.
44. Scott, L., "Intelligence and the Risk of Nuclear War", p765.
45. DiCicco, J.M., "Fear, Loathing and Cracks in Reagan's Mirror Images", p262.
46. If we were to continue to play out the dance on the periodic table of games, we would have to imagine a situation in which each side thinks it's playing a different game. For example, the USSR might think it was in **ShNc** or **DlNc**, while the United States might think

it was in **ShPd** or **AsPd**. Each side is confident of its own peaceable intentions but imagines that the other is playing a more hostile game (Prisoner's Dilemma or Deadlock), creating deadly risks.

47. Even when $a_1 - a_2$ and $r_1 - r_2$ are appreciable, they remain much smaller than a_1 and r_1, for even after a first strike both sides retain huge nuclear capabilities. See, for example, Cimbal, "Revisiting the Nuclear 'War Scare'".

48. It is also possible to add something to each side's payoff under *(Strike, Strike)*, on the principle that the recipient of a full strike prefers to see its opponent similarly struck rather than not. The payoff rankings are then those of Stag Hunt. This makes no difference to the ensuing analysis.

49. In fact our game is a "Bayesian game" of incomplete information. Presenting it and its results this way keeps the analysis more transparent.

50. This situation is more conventionally treated through sequential moves, with notions such as "subgame perfection". One can imagine grave dangers from artificial intelligence handling nuclear responses in such an algorithmic manner. See Lindelauf, *Nuclear Deterrence in the Algorithmic Age*. We prefer to imagine a game with some irrationality explicitly incorporated.

51. Economic data (GDP per capita used for America, A and Russia, R) is from Jutta Bolt, Robert Inklaar, Herman de Jong and Jan Luiten van Zanden, "Maddison Project Database", version 2018. Arsenal data (warheads used for America, a_1 and Russia, r_1) is taken from Max Roser, Bastian Herre, and Joe Hasell, Nuclear Weapons, 2013. In the game, we take these to be normalized by the sums, turning these variables into zero-sum values in the sense that if one side improves their economy or arsenal, the other side feels worse off. Then for Figure 5.12, $a_2 = a_1 - 0.5\,r_1$ and $r_2 = r_1 - 0.5\,a_1$, the amount of your arsenal left untouched by your opponent's first strike if half of that first strike damages your share of warheads.

52. For first-strike advantage – the proportion of one's opponent's arsenal mitigated by a pre-emptive strike – we pinned the values at 0.4 (1945: both), 0.4 (1957), 0.2 (1970, following the move from primarily airborne weapons to ICBMs and SLBMs), 0.2 (1977), 0.2 for a, 0.5 for r (1983, SS20s mostly deployed but not Pershing IIs), 0.6 (1985: both, with Pershing IIs recently and rapidly deployed), 0.6 (1987), 0.4 (1994) and interpolated linearly between these.

53. It would be easy to make $p_A^C \times p_R^C$ more closely follow the more exaggerated movements of the clock by using a sigmoid link function.

Our first-strike advantage timeline matches the actual rather than planned deployment of SS20 and Pershing II, which perhaps explains why the clock anticipates the game. Note that there is some question as to where certain peace, "100%" chance of no nuclear war should be, compared to the minutes to midnight; we here take it to be 15 so that the maxima roughly agree between clock and game.

54. Mastny, "How Able was Able Archer?" p119.
55. Further, we would say that our injection of a totally irrational step – the arbitrary ascription to the other of some likelihood of a first strike – is at least as realistic as any game which relies on rationality.

CHAPTER **6**

Conclusions

T HERE IS ONLY ONE course of history to analyse – yet if the
analysis is to be anything more than reportage then it must
admit alternatives, even if only implicitly through historians' use
of the language of contingency. As we saw in the opening chapter,
endless words can and have been written about this conundrum.
Our aim has been to cut through the debate by seeking *productive*
treatments of alternative history – productive not of fictions but
of machinery with which useful quantifications can be made, and
of a different perspective on historical analysis, offering different
lights and shadows.

We began with dreadnought battleship combat in the First
World War, which we assert is, perhaps uniquely in warfare, well
described by Lanchester's simple mathematical model. We could
then use the techniques of approximate Bayesian computation,
which at its simplest is a matter of challenging prior beliefs with
real data and updating them. The crux was to use this to identify
a situation in which something unlikely happened – in this case,
at the battle of Dogger Bank – and its implications for the later,
much larger action at Jutland. One perspective that emerges is of
the battles' being embedded in the wider context of force planning, that strategic control of the North Sea throughout the war

DOI: 10.1201/9780429488405-6

by the British Grand Fleet was a result of decisions made over the decades before the war, so that the outcomes of individual battles were not so critical, whatever the alleged tactical and material failings. But this is complemented by the dependence on personalities, that what leaders – Beatty, Jellicoe, Hipper, Scheer – do matters and is a function of their characters.

It is, however, unusual to have such a good model, which captures rather than distorts reality. Lanchester's model did not apply to the Second World War Battle of Britain, so we employed a second technique, which takes "bootstrapping" – resampling with replacement, or in this case re-running the battle many times using only the real daily battle data – and modifies it to use more or less of the data with particular characteristics, thereby extending or altering the nature of the simulated battle. This works at a particular, productive point in the spectrum between those who love counterfactual history and those who reject it entirely: it explores alternative possibilities for the battle, but using only data from events that really happened. Above all it is model-free; there are no hidden mechanisms or parameters. The outcome is a clearer sense of what it would have taken for the Germans to have won the battle – "won" in the sense only of triggering a decision to invade England. Again the dependence on personalities was striking. We always aimed to keep our counterfactuals "restrained" – modest, minimal and controlled. Yet when stronger counterfactuals led to German wins, these were not so much "exuberant" fictions as different personalities – compared to Goering and Hitler – with different, wider conceptions of strategy, both military and geopolitical. To argue that the Germans could never have won becomes to argue that such different conceptions could never have been plausible, even under different commanders.

Our third case brought us closer to the information age. Vietnam was the first war in which commanders consciously tried to create new data and use it to prosecute a war, here the "Hamlet Evaluation System" which attempted to measure the state of the country's rural settlements. The war gave rise to the famous

"McNamara fallacy",[1] that only what can be measured should be used in decision-making, which is fallacious because data is a mere shadow of reality which distorts the decision-making process if one succumbs to the illusion that the data *is* reality. Better data, better analytical techniques and more computer power allow us to do better, but never wholly to transcend the fallacy. A historical perspective then offers lessons for the modern age. In practical terms, we can see from a modern analysis of the HES data that an approach to Vietnam which emphasized pacification rather than attrition might have offered an improved outcome for the Americans and their South Vietnamese allies, if implemented early enough. Once again, policies appear to have been largely determined by personalities, again operating within the perceived political realities and conceptions of their time.

Finally we moved to Able Archer, where actors consciously used game theory and its associated notions to avert war in the nuclear age. But limited conceptions were often opposed to the underlying truth. The mere fact that our invented game captures what happened better than those of the time should not lead us to the conceit that we would not replicate this error if we applied our game to current concerns. The higher level lesson is surely always to continue talking to one's opponents, simply to better understand the rules of *their* game – and that remaining lazily or wilfully ignorant of these is unlikely to bring a better outcome, while attempting canny deceptions runs grave risks.

So, in the end, perhaps we have reached the limits and been reminded to be conscious of the limitations of quantitative modelling. Famously, "all models are wrong, but some are useful".[2] This lesson has to be re-learned in each new field in which modelling is applied, lest the sheer power of modern computing seduce an inexperienced user into thinking that modelling is simply a matter of inputting parameters and data into a "black box" to discover *the* answer. In our own collaborations, the engineering in the black box is precisely the first subject of interest, to both mathematicians and historians, with the aim that everyone should understand it. Only then can we observe the light it shines on history – and, increasingly through the twentieth century, on the historical actors, from Jellicoe to McNamara.

Beyond the introduction of some new techniques into historical analysis and the specific lessons of each chapter, what broader lessons have we learned? First is the lesson of interdisciplinary working. Nowadays, no individual can grasp all of human intellectual endeavour, but interdisciplinary science teams have long experience of fusing their expertise so that the team becomes the polymath. Ours is one of the smaller number of teams doing the same across the larger gap between the sciences and the humanities. The only necessities beyond individual expertise are shared interest, humility and a willingness to talk – and listen!

Next, we have learned the danger of hubris. As data and modelling become more available and prevalent, we must not be tempted into a Whiggish view that *now* we have the correct tool for every given question. Current tools, current thinking will be superseded. Human understanding remains beyond available data or models, and artificial intelligence and machine learning too easily remain black boxes, the workings of which we must labour to understand, and which must never be operated blindly. The central idea of the McNamara fallacy[3] should be widely known, and its compass extended to these techniques of the modern age. One of the great dangers of the siloing of human intellectual interaction, of "filter bubbles" and the like, is that people cease working to understand other's positions – and models must never be allowed to stand in the way of this.

Finally, we are struck, despite our intentions, by the extent to which personalities matter once one moves beyond the most restrained and cautious of counterfactual possibilities. If nothing else our examples have enabled some level of unbiased identification of who these people are, and which decisions mattered at the critical juncture and by how much. This, in turn, can shed light on the broader context – doctrines that enable good decision-making, and data useful in supporting it. People together make history as we know it, whether that means making decisions which turn out to be correct and are rewarded, or making wrong decisions – wrong because they were ill-considered or ill-informed, and thus always unlikely to turn out well, or merely because they were unlucky. In the end, everyone has to accept the roll of the dice.

NOTES

1. Originally due to Daniel Yankelovich, as quoted in for example O'Mahony, "Medicine and the McNamara fallacy."
2. As noted earlier, this is the broad argument of Box, *Science and Statistics*.
3. Indeed the McNamara fallacy, in the form that only what can be *measured* is important, morphs easily into the "streetlight effect" referred to earlier, that only what can (currently) be *modelled* is important, and thence into a Whiggish position that current modelling is correct modelling.

Appendix

Some Mathematical Background (with No Equations!)

THIS APPENDIX is *not* a full introduction to the theory of probability and statistics, much less the whole of mathematics. We provide here a series of important concepts that provide useful background to the main book. The reader certainly doesn't need to understand all of this appendix to read the book. Rather, sight of the contents here will provide the reader with a fuller picture of the issues raised. It is also intended as a guide for the perplexed, and especially for those who left maths or science behind at school or college.

There are already plenty of good books[1] that provide a fuller background to the topics raised, and we warmly encourage readers to investigate further. The sections below are loosely tied to the order in which they appear in the main text, and each has its own references, which ultimately move towards the technical or research articles. Our own collaboration works only because the historians have the right to expect the mathematicians to strip every argument back to $1 + 1 = 2$ – and this is good for the mathematicians, too, because it allows the historians to spot and ask important questions, and indeed it is precisely such questions that often sow the seeds of progress.

PRELIMINARY: LARGE AND SMALL NUMBERS

Right, $1 + 1 = 2$; we can all do addition. We can all multiply, as well. However, multiplying large numbers sometimes causes problems. Notice that if I multiply 1,000 by 10,000 to get 10,000,000, what I'm doing is just counting zeros: $3 + 4 = 7$. And this is how scientists handle large numbers, by writing this calculation as $10^3 \times 10^4 = 10^7$, where 10^3 means $10 \times 10 \times 10$, and so on. This power of 10 is called (in very old terminology whose origin is forgotten by most) a "logarithm", or "log" for short. Logs turn multiplication into addition: $1,000 \times 10,000 = 10,000,000$ becomes $\log(1,000) + \log(10,000) = \log(10,000,000)$ or $3 + 4 = 7$. Multiplying by 1 does nothing, just as adding 0 does nothing, so we write $10^0 = 1$ or $\log(1) = 0$.

Many things are measured on a "logarithmic scale". For example, *adding* 1 bel (more commonly called 10 decibels) corresponds to *multiplying* sound energy by 10. If an earthquake measures 1 *more* on the Richter scale[2] than another then it has released 10 *times* as much energy; 2 more and the energy is 100 times greater. Many human perceptions are like this: we perceive multiplied effects as additive, as something more.

We can also deal with small numbers in precisely the same way: $0.001 \times 0.0001 = 0.0000001$. If you're tempted to say that the numbers of zeros – the zeros after the decimal point, that is – no longer simply add, I'd ask you to think of this as $1/1,000 \times 1/10,000 = 1/10,000,000$. The logarithmic version of this is $-3 - 4 = -7$. We write $0.001 = 1/1,000 = 10^{-3}$: the minus signs work nicely because then $1/1,000 \times 1,000 = 1$ becomes $-3 + 3 = 0$.

At the level of precision at which we are working, we are mostly interested in the first significant figure, or even only the relative sizes of the exponent. This is very much how physicists understand the universe – for example, we say that there are something like 10^{23} stars in the universe,[3] give or take a factor of ten either way. Similarly, we need small numbers to handle rare events, such as the "one in a million" event that happens with a probability of 10^{-6}. At this level, it is easy to lose sight of the fact that the *ratio* of probabilities of, say, an event with a probability of a half, a coin toss, and 0.005, the chance of a birthday falling on one of two days, is roughly the same as that between events with probabilities 10^{-7} and 10^{-9}, say. But the ratio in each case is $1/100 = 10^{-2}$ and can be seen from the -2 shift of the log.

PROBABILITY AND SURPRISE

Probability is a vital area for this book, and an area of mathematics which is both widely applicable to the modern world and widely misunderstood. The famous Monty Hall problem,[4] with its prizes of goats and Ferraris, still confuses to this day but is at its heart a simple statement of probability. Quantifying probabilities is a necessary part of this book, as we noted above, and an object which has become core to human civilization,[5] a coin, when "tossed", provides a perfect lens through which to view events.

Of course, the tossed coin is just our cultural way of understanding a binary event. Mathematically this gives rise to what is known as a binomial distribution, which has many attractive properties, not least that it rapidly becomes a bell curve (see next section) once the coin is tossed repeatedly. But to see through this lens more clearly, we need to shift our focus. We've used "logs to base 10", but we could as easily have used another number. Science should not depend on the accident that we happen to have ten fingers. Let's use two and create a log scale for probabilities.

If you correctly call heads on the toss of what you believe to be a fair coin, you're perhaps no more than a little bit surprised and pleased – after all, it was a "50:50" chance; the probability of heads was ½. If you call heads twice, the chance that they both come up is ½ × ½ = ¼, (because we multiply probabilities like this when the two events are independent) and you're a bit more surprised. If you correctly call heads ten times in a row, you're very surprised (and will have to believe either that you got very lucky or that the coin was biased), for the chance of this happening was ½ × … ½ = 1/1,024 or about 10^{-3}. So let's create a log scale for our "surprise" (or sometimes "surprisal"), in which we experience, respectively, 1, 2 and 10 "bits" of surprise in these three cases. This gives us a nice additive scale for the extent of our surprise when unlikely things happen. It is also very close to the log of the "odds" of a bet.

In fact, this idea is at the root of modern information theory. If we want to measure how informative unlikely independent events are, then we want their information to *add*, yet their probabilities *multiply*. The only way to achieve this is to use a logarithmic scale, and this is the basis of Shannon's "information entropy".

The same idea enables us to discriminate between the accurate and the not-so-accurate in competitions which require forecasts in the form of probabilities of future events. Competitors are then scored after the events do or don't happen and the most successful have become known as "superforecasters". There are different ways of scoring, but one of these, the "log score", measures, over many events, the average extent to which the forecaster was surprised, in the above sense. The "Good Judgment Project" of Mellers and Tetlock, on the other hand, uses a different, non-logarithmic score, which instead measures the square of the difference between the forecast and the event.[6]

We'll use the notion of surprise to ask how surprising a forecaster might find various quantitative results. Of course, unlikely things *do* happen – but not very often (by definition). To argue that a real event was highly improbable, like the outcome of the battle of Dogger Bank, requires a strong argument. On the other hand, counterfactual events which would have been very unlikely in reality, like our counterfactual Battle of Britain outcomes, effectively measure the strength of the counterfactual conditions necessary to make such outcomes likely.

BAYES' IDEA

Thomas Bayes' idea has become an enormously popular way of thinking about data in recent years, an alternative to the preceding century's tradition of statistics. In traditional statistics, there's usually a "null" hypothesis, an idea to be challenged by the new data, of "nothing happening here". The problem with this is subtle, and it comes down to what we choose to test; there are an infinite number of possible questions and hypotheses and our selection betrays a bias, often an unintentional one. In Bayesian statistics, by contrast, one always begins with some "prior" view, a distribution representing our initial biases, and then updates it using the new data.

At its simplest, Bayes' idea begins with different ways of writing the probability of two things being true at once, the data and an explanatory model – symptoms and a disease, say. This could be written *either* as the probability that the patient has the disease, multiplied by the probability that someone who has the disease has these symptoms, *or* as the probability that the patient has the symptoms, multiplied by the probability that someone who has the symptoms has the disease.

The distinction is subtle but crucial. Many hypochondriacs over the years have researched their symptoms and concluded that they have a specific disease which produces them.[7] But what they have usually noticed is that almost all people with that disease have those symptoms, which is not the same thing at all as saying that almost all people with the symptoms have the disease. To tease out the distinction, we use the previous paragraph. We estimate the separate probabilities of the disease and the symptom simply as the number of such people in the population, divided by the total population. Then the probability of the disease, in someone with the symptoms, is the probability of the symptoms, in someone with the disease, multiplied by the number of people who have the disease, and divided by the number of people who have the symptoms.

For example, suppose I have a symptom which is typically produced by a certain cancer. Do I have cancer? The probability of the symptom, given the disease, is almost 100%. But lots of other diseases also produce this symptom. The probability that I have the cancer is, roughly, its frequency in the population divided by that of the symptom.

In the more general Bayesian approach I begin, before observing the symptom, with my own "prior" estimate of the probability that I have cancer. Suppose 0.02% – 2 in 10,000 – of the population have the cancer. Then I observe the symptom. The ("posterior") probability that I have cancer after observing the symptom is the prior, 0.02%, multiplied by a factor which is the probability that I would have the symptom if I had cancer (the "likelihood", nearly 100%), and then divided by the probability that I have the symptom (the "evidence"). If 1% of the population have the symptom, then this factor is about 100% divided by 1%, or 100. My posterior is then my prior multiplied by this factor, or about $100 \times 0.02\% = 2\%$. In a population of 10,000, from being a random person with no more than the base chance of 0.02% of having cancer, I've now become one of the 100 people who have the symptom, of whom 2 have cancer – so my posterior probability is 2%. So things look a little bleaker, sure, but the hypochondriac is still in error by a factor of 50.

AS EASY AS ABC

The example above was a simple way of viewing probabilities. For 200 years, it seemed to be no more than that – a philosophical discussion, not a practical one. The curious history of Bayes and his idea merely reinforces this notion, that it is a matter only of world view and outlook. The problem is that to make this mode of reasoning practical for most purposes, computers are needed. Here are two reasons why this is so.

First, in our hypothetical cancer example above, our prior was just a single number, 0.02%. In reality, there would be a great deal of uncertainty about our personal base probability of cancer – we might say "somewhere between 1 in 10,000 and 3 in 10,000", or "somewhere between 0.01% and 0.03%", or "0.02% plus or minus 0.01%", according to age, sex, various lifestyle parameters, and so on. Most priors in reality have some such degree of uncertainty, and so take the form of distributions of probabilities – something like a bell curve for the probability itself – dependent on underlying parameters. Then one has to sample – to "try out" – many different individual prior probabilities, "sampling" from the prior distribution.

Secondly, calculating the likelihood – the probability of the data given a model and the values of its parameters – can be a complicated business. Standard Bayesian reasoning requires us not only to be able to do this but also to be able to write down neatly how the answer depends on the parameters. Often this is impossible. Further, as we shall see below, we shall have to calculate it many times. But in the last 20 years, a set of techniques has been developed which uses computer power to solve these problems. This practice has become known as "Bayesian computation without likelihoods", or Approximate Bayesian Computation (ABC).[8]

To implement ABC, we need data and a model – that is, some results and some mechanistic idea of what is happening. For clarity, let us pick up the concepts from Chapter 2. For the clash at Dogger Bank, we have the results – shots fired, hits scored, ships lost – and we have a model, the Lanchester model. We want to ask questions of the combat such as "how likely was it that the Royal Navy would lose a ship at Dogger Bank?" Our priors are now buried much deeper in the model, in the form of rates of fire, vulnerability to fatal hits, and so on, for which we had reasonable estimates from other historical sources. What is interesting is that some of these priors are considered accurate (which is not surprising – naval officers were very interested in hitting targets) but others are not (how likely was a plunging shot to ignite a magazine?) and we can reflect this with broader, more uncertain distributions for our priors.

To proceed we then have to compute an adjustment factor – recall that this is the likelihood (the probability of the data, given the model), divided by the evidence (the probability of the data), and that is the key goal of ABC. The numerator, the probability of the data emerging from the model, we *can* compute, even if this requires some effort: it's what the model is for, the result from a simulation of the battle. But the denominator is a problem, because we simply don't know the "probability of the data"; we only have our model. What we can do instead is to write the probability of the data as the sum of its probabilities in all possible variations of our model – that is, with all possible values of the parameters. But recall our second problem, that to compute the probability of the data given the model can be hard work. Yet now we have to do this many times, in order to add up the results for all possible values of the parameters. This is tremendously computationally demanding but, with some clever tricks, feasible. This is the core of Bayesian computation without likelihoods, which really only means "without neat mathematical expressions for likelihoods" – which is the mathematicians' problem, not yours. Once it's solved, we write the posterior as the product of the prior and our adjustment factor.

So ABC has three elements: some parameters, a model and some data (both simulation outputs of the model, and what actually happened). We have a "prior" idea of the parameters, feed them into the model, and the model produces some simulated results. We compare these with the data and adjust the parameters using our computed likelihood ratio. And do it again, and again, and again.... The goal is to repeat the procedure so many times that we can be quite sure (first) that the outcome has no memory of which priors we started with, and (secondly, and harder) that we've explored all possible values of all the parameters. We end up with our final posteriors, the probability distribution of whatever summary statistic we're interested in. In Chapter 2, this was the number of ships sunk, say, or the number of hits; in medicine, it might be the effect of a new drug as a function of the dosing regimen.

If this all works nicely, then we have a very good method for adjusting our ideas about the regimen for the drug, or the operational parameters of a weapons system. But suppose it doesn't quite work, in that the true outcome looks to have been highly unlikely. This is especially problematic when, as in history, we only have one real data point. Then either the model is wrong, or our prior estimates of the parameters are very far – unworkably – wrong, or the true outcome really was unlikely. This is where we emerge, spluttering, from the mathematical deep end, because it's really the historical sources that justify our confidence in the model and the parameters. The British really were lucky at Dogger Bank.

BOOTSTRAPPING AND THE BELL CURVE

One can only imagine that the original deviser of problems in probability was an avid collector of urns with an unusual obsession with coloured balls. At the risk of perpetuating this meme, we ask you to suppose you have an urn with 50 blue balls and 50 red balls. You take a ball out, then do it again, and again, 100 times in total. Rather obviously you now have 50 blue and 50 red balls. This can be thought of as having all of the data, the draws, from the population – you now have everything that was in the urn.

But now suppose you take a ball out, note its colour, *put it back,* (mix them well,) and take out another one, and again do this 100 times. This is like tossing a coin 100 times: 100 times over, you picked out a blue ball with probability ½, just as when you tossed a coin it came up heads with probability ½. And just as with the coin, your 100 colours will not now be divided blue:red as 50:50. They could be anywhere between 100:0 and 0:100, although those extremes would surprise you very much; they're much more likely to be somewhere between 60:40 and 40:60. In fact it's an example of one of the most fundamental theorems of statistics that, if you did this procedure many times, the distribution of results would look like a "bell curve" centred at 50:50. Technically, a bell curve is what's called a "Gaussian" or "normal" distribution, and it has a number of nice properties. A typical amount of variation on a bell curve – half the width of the "shoulder", in a way one can make precise – is called a "standard deviation", usually written as a Greek sigma. Unlikely events in a bell curve are often referred to by how many sigmas from the centre they lie. For our example above, sigma is about 5: just under 70% of results are somewhere between 55:45 and 45:55.

Bootstrapping takes this idea and uses it to guess what the population might look like just from the draw of the balls we have.[9] Perhaps our imaginary urn collector won't let us look in their urn. Or say the urn had many more than 100 balls, but someone took it away from us when we had drawn 50 blue balls and 50 red balls. Then we do not know the properties of the other balls in the urn, even if we managed to get a ball of every available colour. We can still work with this. Our bootstrapped data (produced using the procedure of the last paragraph) are giving us some idea of what the proportions in the urn's larger original population should look like. Of course, the bootstrap can never tell us about really unusual balls, if indeed there are any, whose properties were never observed among that first 100. There might be balls of other colours, but there are probably at most a few of them.

Suppose we had drawn just 20 balls before the urn was taken away, and these were, unusually, 15 blue and 5 red. We then have less data and our idea of the original population would be less precise. This would then be reflected in the average estimate made by the bootstrap, which would be that the urn overall contains 3 blue balls to every red. But conducting the bootstrap procedure with just these 20 balls would result in a highly variable distribution, a much broader bell curve, much more so than if we'd had 100 balls – essentially the bootstrap knows and displays its own inaccuracy.

Crucially, though, the bootstrap represents the best guess we can make about the population using only the data we actually have, rather than outside ideas about what the original population might have been like. In that sense, what might be thought of as caveats are instead virtues, as they make it clear just how far we can go with the data as it is. The blessing of the bootstrap is that although we can't necessarily seize the urn and look inside or change the rules, we can ask the mysterious deviser of problems and collector of urns to do it again. And again. And again. In this way, we can build up a statistical picture of increasing accuracy.

WEIGHTED BOOTSTRAPPING AND COUNTERFACTUALS

Or perhaps we can change the rules a bit. Although the urn collector won't let us break open the urn, we can search for different balls – heavier ones, larger ones, whatever. Perhaps there's something slightly different about the "feel" of the red balls that means they are drawn preferentially. Any of these changes will lead to us biasing the draws from the urn. We do our 100-draw-with-replacement experiment and do it many times. How big was our bias? We plot the distribution of the results, and observe that its peak is at around 30 blue and 70 red balls, 30:70. We've already said that sigma is 5 in this case, so we call this shift of the centre from 50 to 70 a "four sigma" shift, since $20 = 4 \times 5$.

This is what we did in our weighted bootstrap in Chapter 3. The balls were replaced by the days of the Battle of Britain in 1940, and the colour by that day's sortie and loss numbers, targets, weather and so on. The biasing was towards particular targets, and was counterfactual in that it differed from the targeting that actually occurred. (We also sometimes ran an extended battle with rather more than the actual number of days.) Each choice of bias, each choice of counterfactual, shifted the bell curve by a number of sigmas.

Recall that our "Counterfactual Hitler" (and German war strategy; call it "CH") and "Counterfactual Goering" (and the Luftwaffe staff; call it "CG") each produced shifts of about three sigmas. A three-sigma event, on a bell curve, has a probability of approximately 1 in 1000, about 0.001, or about 10 bits of surprise. Combining the two gives a six-sigma event, with a probability of about one in a billion, 10^{-9}, or 30 bits of surprise.

As noted earlier, this gives a natural way to classify the size of a counterfactual. *In the battle as it actually occurred*, for the Germans to have got the results of either CH or CG would have been extremely unlikely, highly (10-bits-)surprising. To have got them both would have been astonishing, extremely (30-bits-)surprising.

But in the counterfactual scenarios, such results would have been typical – not surprising at all. The "size" of the counterfactual is how surprising, in its absence, its typical outcome would be.

Notice that doubling the number of sigmas – going twice as far out into the tail of the distribution – *more* than doubled the surprise. This is because the bell curve is a "light-tailed" distribution: extreme cases are exceedingly rare. The distribution of human height is an example: there are no 3 m or 10′ tall people. Bootstrapping gives a bell curve, and extreme results are exceedingly rare. Other processes can produce other kinds of distribution which are "heavy-tailed" – extreme events are rare but not *that* rare, and in which surprise *less* than doubles at twice as far out. Separating light from heavy tailed is the "exponential distribution", in which the surprise is proportional to the distance from typicality.

VIETNAM AND THE HAMLET EVALUATION SYSTEM

Vietnam brings us forward to the era of big data, collected and analysed in near real time with the intention that it could be used to inform operational and policy decisions. In Dogger Bank and Jutland, there were very few real data points. In the Battle of Britain, we had a sequence of data points at increasing times – a "time series", although its progression over time was not our main focus, and so we were not using standard time-series techniques. But in Vietnam we have a huge number of data points, with multiple attributes for every hamlet in South Vietnam, developing over time.[10] In statistics, these are known as "panel data", originally from "panels" of people assembled so that the same group could be observed at multiple times (in "longitudinal studies").

As with much big data, its overwhelming characteristic is that, as gathered, it's *messy* – a problem caused and compounded, in this case, by the urgency with which it was collected and used. By far our main task was cleaning it up. There was inconsistent coding and naming. Approximately 40% of locations were incorrectly recorded. Documentation was erratic. Then, when cleaned, it's clear that a great deal of data is simply *missing* – for example, only 7% of hamlet-level data was recorded for 1967–1969. Missingness is an important topic in big data, but one cannot simply interpolate and impute missing data points, because the missingness may be systematic. (Picture the poor guy sent out to check up on the state of a hamlet he believes is controlled by the Vietcong. Is he really likely to go there and ask who's in charge? – as likely as if he believed the hamlet were friendly?) Further, much of the contextual data was spread out over multiple irreconcilable data sets. This is a particular problem for investigating causation, which often requires an understanding of the wider situation, observed consistently over time – which it rarely is, and wasn't in this case.

Harrumph. Having said all that, and as we saw in the chapter, in the end there was some worth for the historian in the data. Units that were said to have had bad behaviour tended to be found in hamlets that were unusually poorly secured, and places that were more under attack (for example, during the Tet Offensive) had worse outcomes during the times they were under attack. Regions under the control of the USMC tended to do better, and were more likely to be conducting pacification policies. So the central counterfactual, of how the allies might have fared with a greater emphasis on pacification, is somewhat supported by the HES data.

But there was no simple "black box" statistical technique that could discover this; the data were simply too messy to support it. Perhaps that's the final lesson, so typical of working with messy big data of which, since it's from the past, one has no control of the collection. Only full attention from a multidisciplinary team can reveal its hidden truths – messy, incomplete, revealing itself partially and in glimpses, but important for all that.

GAME THEORY

Chapter 5 is a little different from the others, in that we introduced the mathematics of "game theory" as we went along. It does, however, resemble Chapter 4 in that state actors were attempting to use mathematical analysis of what they thought of as data in real time to inform policy decisions and defence strategy. But the reader may feel a little more sceptical of its worth than of the methods in the other chapters, and they would be quite right to do so.

The problem is evident. Just as with economics' reliance on "rational actors", now tempered by the new field of "behavioural economics", game theory has to make assumptions about "theory of mind", the infinite regress of "I think", "I think you think", "I think you think I think..." that is inherent in human dealings – and it never quite encompasses all of this.

As we introduced it, in a two-player game[11] each party knows how the other values each possible outcome, or at least how they are ranked, and assumes that the other acts to optimize this. The clearest situation is when there is a "dominant strategy", when "I am going to do this, and I don't care what you do" – because, for every possible choice you can make, the same course of action is best for me. By contrast, the other two notions of optimality are less satisfactory. Pareto-optimality says "let's do this – it makes me better off, and you will be no worse off". An obvious objection is that you might take perverse satisfaction in preventing my doing better, but this can be factored into calculation of the payoffs. The greater problem is that I really don't care whether or not you are worse off. If, in Prisoner's Dilemma, we are both *Quiet*, I clearly do better if I *Talk*. The Nash equilibrium, too, has its problems: it is really no more than a state of feeling trapped, in that I can do no better by changing strategy.

There are many ways of augmenting game theory with more true-to-life human behaviours and uncertainties. At the mathematical end, in a Bayesian game (a "game with imperfect information"), I don't know your payoffs, but rather have an idea of the probabilities of your different possible payoffs, and essentially play a normal game with these in mind. In fact our Nuclear Stand-Off game was of this form, but presented differently. "Behavioural game theory" incorporates many such uncertainties, and often adds the twist of using the experimental outcomes of games to reverse-engineer the players' true strategies and objectives.

But in our view the reader would at last be correct to judge that the whole subject is not yet fit to predict people's choices – and certainly not with sufficient certainty to be relied on for the aversion of thermonuclear war. Our game, in which one actor (completely arbitrarily and irrationally) imputes to the other a probability of a nuclear first strike, and then computes from their own current and possible future state the value of a pre-emptive strike, is merely one illustrative way to cut through the thicket.

LANCHESTER'S MODELS

Finally, and out of order, we come to Lanchester's models. For those who have a little calculus, somewhere between advanced school and lower university level, there are many introductions available.[12]

Suppose you want to know who will win a battle. Clearly surprise, shock, exploitation, command, morale and many other factors are of crucial importance. However, suppose that all we know are the *numbers* on each side, and some measure of how *effective* each side's units are. A natural supposition might be that the side that wins is that with the larger value of numbers multiplied by effectiveness. Lanchester shows that there are many circumstances in which this is indeed correct – a hand-to-hand battle and a battle of unaimed shots in the sense of fire aimed at an enemy force but not at specific targets within it (artillery or archery, for instance) are the main examples.

But Lanchester is concerned with very unusual circumstances, which he is trying in 1913 to anticipate, of long-range aimed weapons, so that every unit on each side can find an unambiguous target. In this circumstance, each side causes damage in proportion to its numbers. What is the battle-winning combination of numbers and effectiveness that results? This is where calculus – which handles rates of change, small differences, and their summing to give longer-term totals – comes in.

Suppose Red fights Blue. In this "aimed fire" model, the instantaneous casualty exchange ratio, the ratio of Red losses to Blue losses over a short period, is proportional to Blue numbers divided by Red numbers. Thus, Red numbers multiplied by Red losses is proportional to the same for Blue. Calculus enables us to sum these up: if Red starts with N units and loses them one by one, what matters is N + N −1 + N −2 + ···· + 2 + 1, which is (roughly, when N is large) ½ N².

It is this summation which leads to the "N-squared law": that the side with the greater product of effectiveness multiplied by the *square* of its numbers will be the winner. In marginal terms, a 10% advantage in numbers is equivalent to a 20% advantage in effectiveness. The result would be that a commander who trusted Lanchester equations would be unwilling to give battle if even slightly outnumbered, as a decisive defeat was assured.

Deitchman's model for insurgencies, referred to in Chapter 4, modifies the square law in the following way. Suppose Red is an insurgent, so that Blue finds it hard to identify Red targets. In the model, Blue's aimed fire is diluted, multiplied by a density factor which is the Red force divided by a much larger "dilution parameter". The outcome is that, instead of its standard square-law fighting strength of effectiveness times numbers squared, Red has a fighting strength which is its numbers, multiplied by its effectiveness, and then multiplied by double the dilution parameter. The doubling is a consequence of how the summing effect works in this model, and Red can only make limited improvements to its units' effectiveness, but the dilution parameter is wholly under its control. To increase its fighting strength enormously, all Red has to do is melt away.

NOTES

1. Silver, *The Signal and the Noise*; Pinker, *Rationality*; Spiegelhalter, *The Art of Statistics*; Pearl and Mackenzie, *The Book of Why*; and finally Bailer-Jones, *Practical Bayesian Inference* for the scientifically educated and mathematically-minded.
2. The Richter scale itself is most commonly used by the media these days, but logarithmic scales are still used by scientists for the various "moment magnitude scales".

3. https://www.esa.int/Science_Exploration/Space_Science/Herschel/How_many_stars_are_there_in_the_Universe

4. There are three doors, behind two of which are goats while behind the third is a Ferrari. The host (who knows where the Ferrari is) says: "pick a door". You do. He opens a different door, revealing a goat (which he knows is there). He says, "do you want to change your choice?" Should you?

5. https://www.bbc.co.uk/programmes/b00qm8zg

6. Tetlock and Gardner, *Superforecasting*; https://www.gjopen.com/

7. Perhaps everything except housemaid's knee, as the central character in Jerome K. Jerome's *Three Men in a Boat* worries.

8. Marjoram *et al.*, "Markov chain Monte Carlo without likelihoods."

9. Efron and Tibshirani, *An Introduction to the Bootstrap.*

10. https://www.archives.gov/research/military/vietnam-war/electronic-data-files

11. A classic nontechnical introduction is Davis, *Game Theory.*

12. A brief introduction by one of the authors is MacKay, "Lanchester combat models".

Bibliography

Adamsky, Dmitry D. "The 1983 Nuclear Crisis, Lessons for Deterrence Theory and Practice." *Journal of Strategic Studies* 36, no. 1 (2013): 4–41.

Andrade, Dale. "Westmoreland Was Right: Learning the Wrong Lessons from the Vietnam War." *Small Wars and Insurgencies* 19, no. 2 (2008): 145–181.

Andrew, Christopher M. and Oleg Gordievsky. *Comrade Kryuchkov's Instructions: Top Secret Files on KGB Foreign Operations, 1975–1985.* Palo Alto, CA: Stanford University Press, 1993.

Anon., *German Plans for the Invasion of England: Operation Sealion, 1940,* Declassified Central Intelligence Agency report, published by Digital Publications 2017.

Bailer-Jones, Coryn A. L. *Practical Bayesian Inference.* Cambridge, UK: Cambridge University Press, 2017.

Barrass, Gordon. "Able Archer 83: What Were the Soviets Thinking?" *Survival* 58, no. 6 (2016): 7–30.

Baudry, A. *The Naval Battle: Studies of Tactical Factors.* London: Rees, 1914.

Ben-Menahem, Yemima. "Historical Contingency." *Ratio* 10, no. 2 (1997): 99–107.

Berlin, Isaiah. *Historical Inevitability.* Oxford: University Press, 1954.

Berlin, Isaiah. *The Hedgehog and the Fox: an Essay on Tolstoy's View of History.* London: Weidenfeld and Nicholson, 1953.

Bindoff, S. T. *Tudor England.* Pelican History of England, vol. 5. London: Penguin, 1964.

Bolt, Jutta, Robert Inklaar, Herman de Jong and Jan Luiten van Zanden. *Maddison Project Database*, version 2018, 2018.

Bonneuil, N. "The Mathematics of Time in History." *History and Theory* 49, no. 4 (2010): 28–46.

Box, George E.P. "Science and Statistics." *Journal of the American Statistical Association* 71, no. 356 (1976): 791–799.

Brams, Steven J. "Theory of Moves." *American Scientist* 81, no. 6 (1993): 562–570.

Brams, Steven J. *Superpower Games*. New Haven, CT: Yale University Press, 1985.

Braudel, Fernand. *A History of Civilizations*. New York, NY: Penguin Books, 1995.

Bruns, Bryan Randolph. "Names for Games: Locating 2× 2 Games." *Games* 6, no. 4 (2015): 495–520.

Bruns, Bryan. "Escaping Prisoner's Dilemmas: From Discord to Harmony in the Landscape of 2x2 Games." *arXiv preprint, arXiv:1206.1880* (2012).

Brynen, Rex. "Virtual Paradox: How Digital War Has Reinvigorated Analogue Wargaming." *Digital War* 1 (2020): 138–143.

Buchan, John. *Oliver Cromwell*. London: Reprint Society, 1941.

Buchan, John. *The Causal and the Casual in History*, the 1929 Rede Lecture, University of Cambridge. Cambridge, UK: Cambridge University Press, 1929.

Bungay, Stephen. *The Most Dangerous Enemy: a History of the Battle of Britain*. London: Aurum, 2010.

Bunzl, Martin. "Counterfactual History: A User's Guide." *American Historical Review* 109, no. 3 (2004): 845–858.

Burriss, Larry. "Slouching Toward Nuclear War: Coorientation and NATO Exercise Able Archer 83." *International Journal of Intelligence, Security, and Public Affairs* 21, no. 3 (2019): 219–250.

Buzzanco, Robert. *Masters of War: Military Dissent and Politics in the Vietnam Era*. Cambridge, UK: Cambridge University Press, 1996.

Campbell, John. *Jutland: an Analysis of the Fighting*. London: Conway, 1998.

Capoccia, Giovanni and R. Daniel Kelemen. "The Study of Critical Junctures: Theory, Narrative, and Counterfactuals in Historical Institutionalism." *World Politics* 59, no. 3 (2007): 341–369.

Carlson, Lisa J. and Raymond Dacey. "Sequential Analysis of Deterrence Games With a Declining Status Quo." *Conflict Management and Peace Science* 23, no. 2 (2006): 181–198.

Carlyle, Thomas. *On Heroes, Hero-Worship and the Heroic in History*. London: James Fraser, 1841.

Carr, E. H. *What Is History?* The George Macaulay Trevelyan lectures delivered in the University of Cambridge January–March 1961. London: Macmillan & Co, 1962.

Chase, J. V. "A Mathematical Investigation of the Effect of Superiority of Force in Combats upon the Sea," unpublished secret paper, 1902, reprinted in Appendix C of Bradley A. Fiske, *The Navy as a Fighting Machine* (New York, 1916), reissued in the *Classics of Sea Power* series, Annapolis, MD: Naval Institute Press, 1988.

Cimbala, Stephen J. "Revisiting the Nuclear 'War Scare' of 1983: Lessons Retro- and Prospectively." *Journal of Slavic Military Studies* 27, no. 2 (2014): 234–253.

Clifford, Clark M. "A Viet Nam Reappraisal: The Personal History of One Man's View and How It Evolved." *Foreign Affairs* 47 (1969): 609–617.

Collins, Randall. "The Uses of Counter-Factual History. Can There Be a Theory of Historical Turning Points?" *Amsterdams Sociologisch Tijdschrift* 31, no. 3 (2004): 275–296.

Collins, Randall. "Turning Points, Bottlenecks, and the Fallacies of Counterfactual History." *Sociological Forum* 22, no. 3 (2007): 247–269.

Connable, Ben. *Embracing the Fog of War: Assessment and Metrics in Counterinsurgency.* Santa Monica, CA: RAND Corporation, 2012.

Connable, Ben, Michael J. McNerney, William Marcellino, Aaron B. Frank, Henry Hargrove, Marek N. Posard, S. Rebecca Zimmerman, Natasha Lander, Jasen J. Castillo, James Sladden, Anika Binnendijk, Elizabeth M. Bartels, Abby Doll, Rachel Tecott, Benjamin J. Fernandes, Niklas Helwig, Giacomo Persi Paoli, Krystyna Marcinek and Paul Cornish. *Will to Fight: Returning to the Human Fundamentals of War.* Santa Monica, CA: RAND Corporation, 2019.

Cowley, Robert ed. *What If? Military Historians Imagine What Might Have Been.* New York, NY: Putnam, 1999.

Cox, Richard, ed. *Operation Sea Lion.* London: Futura, 1974.

Csilléry, Katalin, Michael GB Blum, Oscar E. Gaggiotti and Olivier François. "Approximate Bayesian Computation (ABC) in Practice." *Trends in Ecology & Evolution* 25, no. 7 (2010): 410–418.

Davis, Morton D. *Game Theory: A Nontechnical Introduction.* Minneola, NY: Dover, 1997.

Davis, William C. *Duel Between the First Ironclads.* New York, NY: Doubleday, 1975.

Deitchman, S. J. "A Lanchester Model of Guerrilla Warfare." *Operations Research* 10, no. 6 (1962): 818–827.

Deitchman, Seymour J. *Limited War and American Defense Policy.* Cambridge, MA: MIT Press, 1964.

Deitchman, S. J., B. Sheldon and Y Wong. "Military Operations Research Society (MORS) Oral History Project, Interview of Mr Seymour J. Deitchman." *Military Operations Research*, 15, no. 2 (2010): 61–106.

DiCicco, Jonathan M. "Fear, Loathing, and Cracks in Reagan's Mirror Images: Able Archer 83 and an American First Step Toward Rapprochement in the Cold War." *Foreign Policy Analysis* 7, no. 3 (2011): 253–274.

Douglass, Rex W. *The Digital Vietnam War: Big Data from over a Decade of Combat in Southeast Asia, v1.0.* (2011). https://esoc.princeton.edu/data/administrative-boundaries-southeast-asia-countries.

Downing, Taylor. *1983: The World at the Brink.* London: Little, Brown, 2018.

Efron, Bradley and Robert J. Tibshirani. *An Introduction to the Bootstrap.* Boca Raton, FL: Chapman & Hall/CRC, 1994.

Ellsberg, Daniel. *The Doomsday Machine: Confessions of a Nuclear War Planner.* London: Bloomsbury, 2017.

Enthoven, Alain C. and K. Wayne Smith. *How Much Is Enough? Shaping the Defence Programme, 1961–1969.* New York, NY: Harper Colophon, 1972.

Epstein, Joshua M. "Why Model?" *Journal of Artificial Societies and Social Simulation* 11, no. 4 (2008): 12.

Evans, Richard J. *Altered Pasts: Counterfactuals in History.* Waltham, MA: Brandeis University Press, 2014.

Fearon, James D. "Counterfactuals and Hypothesis Testing in Political Science." *World Politics* 43, no. 2 (1991): 169–195.

Ferguson, Niall. "Virtual History: Towards a 'chaotic' Theory of the Past." In *Virtual History: Alternatives and Counterfactuals.* London: Picador, 1997, pp. 1–90.

Fisher, H. A. L. "Modern Historians and Their Methods." *Fortnightly Review* 56, no. 336 (1894): 803–816.

Fiske, Cdr B. A. "American Naval Policy." USNI Prize Essay. *Proceedings of the United States Naval Institute* 31 (1905): 1–80.

Fogel, Robert W. *Railroads and American Economic Growth: Essays in Econometric History.* Baltimore, MD: Johns Hopkins Press, 1964.

Forester, Cecil Scott. "If Hitler Had Invaded England". In *Gold from Crete: Short Stories.* London: Michael Joseph, 1971.

Freedman, David H. "Why Scientific Studies Are so Often Wrong: the Streetlight Effect." *Discover Magazine* 26 (2010): 1–4.

Fry, Stephen. *Making History.* London: Hutchinson, 1996.

Fuller, J. F. C (anon.). "The Principles of War, With Reference to the Campaigns of 1914–15." *Journal of the Royal United Services Institution* 61 (1916): 1–40.

Gelb, Leslie H. and Richard K. Betts. *The Irony of Vietnam: The System Worked.* Washington, DC: Brookings Institution Press, 2016.

Gleick, James. *Chaos.* London: Heinemann, 1988.

Goforth, David and David Robinson. *Topology of 2×2 Games.* Abingdon: Routledge, 2004.

Goode, S M. "A historical basis for force requirements in counter insurgency", *Parameters* (Winter 2009–2010): 45–57.

Gordon, Andrew. *The Rules of the Game: Jutland and British Naval Command.* London: John Murray, 1996.

Daddis, Gregory A. "Choosing Progress: Evaluating the 'Salesmanship' of the Vietnam War in 1967." In *Assessing War: The Challenge of Measuring Success and Failure*, edited by Leo J Blanken, Hy Rothstein and Jason J Lepore, 185. Washington, DC: Georgetown University Press, 2015.

Grinnell-Milne, Duncan. *The Silent Victory.* London: Bodley Head, 1958.

Harrison, Harry. *Tunnel Through the Deeps.* New York, NY: G. P. Putnam's Sons, 1972.

Hennessy, Michael A. *Strategy in Vietnam: The Marines and Revolutionary Warfare in I Corps, 1965–1972.*

Herzog, Todd. "'What Shall the History Books Read?' The Debate Over *Inglourious Basterds* and the Limits of Representation." Ch.13 of Von Dassanowsky, Robert ed. *Quentin Tarantino and* Inglourious Basterds: *a Manipulation of Metacinema.* London: Bloomsbury, 2012.

Hess, Gary R. *Vietnam: Explaining America's Lost War.* Chichester: Wiley & Sons, 2015.

Holwell, Sue and Peter Checkland. "An Information System Won the War." *IEE Proceedings-Software* 145, no. 4 (1998): 95–99.

Horwood, Ian, Niall MacKay and Christopher Price. "Concentration and Asymmetry in Air Combat: Lessons for the Defensive Employment of Air Power." *Air Power Review* 17, no. 2 (2014): 68–91.

Hunt, Tristram. Pasting over the past. *The Guardian*, 7 April 2004. London.

James, T. C. G. *The Battle of Britain.* London: Cass, 2000.

Johnson, Ian R. and Niall J. MacKay. "Lanchester Models and the Battle of Britain." *Naval Research Logistics* 58, no. 3 (2011): 210–222.

Jones, Nate. "Countdown to Declassification: Finding Answers to a 1983 Nuclear War Scare." *Bulletin of the Atomic Scientists* 69, no. 6 (2013): 47–57.

Jones, Nate. The Able Archer 83 Sourcebook. NSA project archive, https://nsarchive.gwu.edu/project/able-archer-83-sourcebook

Kalyvas, Stathis N. and M. A. Kocher. "The Dynamics of Violence in Vietnam: An Analysis of the Hamlet Evaluation System (HES)." *Journal of Peace Research* 46, no. 3 (2009): 335–355.

Kaplan, Fred. *The Wizards of Armageddon*. Palo Alto, CA: Stanford University Press, 1991.

Kinnard, Douglas. *The War Managers: American Generals Reflect on Vietnam*. Hanover, NH: Da Capo, 1991.

Krakauer, D. C. "The Quest for Patterns in Metahistory." *Santa Fe Institute Bulletin* 22, no. 1 (2007): 32–39.

Krepinevich, Andrew F Jr. *The Army and Vietnam*. Baltimore, MD: Johns Hopkins University Press, 1988.

Kwon, Heonik. "Anatomy of US and South Korean Massacres in the Vietnamese Year of the Monkey, 1968". *Asia Pacific Journal, Japanese Focus*, 5, no. 6 (4 June 2007), https://apjjf.org/-Heonik-Kwon/2451/article.html, Accessed 4 January 2023.

Lanchester, Frederick William. *Aircraft in Warfare: The Dawn of the Fourth Arm*. London: Constable, 1916., based on articles in *Engineering* magazine of 1913-14.

Larsen, Stanley Robert and James Lawton Collins. *Vietnam Studies, Allied Participation in Vietnam*. Washington, DC: Department of the Army, 2005.

Lindelauf, Roy. "Nuclear Deterrence in the Algorithmic Age: Game Theory Revisited." In *Netherlands Annual Review of Military Studies*. Asser: the Hague, 2020, pp. 421–436, Ch22.

Luong, Dien. "It's Time for South Korea to Acknowledge its Atrocities in Vietnam", *Foreign Policy* (30 December 2022), https://foreignpolicy.com/2022/12/30/vietnam-war-south-korea-massacres-history-diplomacy/#:~:text=For%20South%20Korea%2C%20what%20is,memorial%20projects%20at%20massacre%20sites., Accessed 4 January 2023.

MacKay, N. J. "Lanchester Combat Models", *Mathematics Today* 42 (2006): 170–173; https://arxiv.org/pdf/math/0606300.pdf

MacKay, Niall and Christopher Price. "Safety in Numbers: Ideas of Concentration in Royal Air Force Fighter Defence from Lanchester to the Battle of Britain." *History* 96, no. 323 (2011): 304–325.

MacKay, Niall, Chris Price and Jamie Wood. "Dogger Bank: Weighing the Fog of War." *Significance* 14, no. 3 (2017): 14–19.

MacKay, Niall, Christopher Price and A. Jamie Wood. "Weighing the Fog of War: Illustrating the Power of Bayesian Methods for Historical Analysis Through the Battle of the Dogger Bank." *Historical Methods* 49, no. 2 (2016): 80–91.

Macksey, Kenneth. *Invasion: The German Invasion of England, July 1940.* London: Greenhill, 1980.

Manchanda, Arnav. "When Truth Is Stranger than Fiction: The Able Archer Incident." *Cold War History* 9, no. 1 (2009): 111–133.

Marder, Arthur J. ed. *Fear God and Dread Nought: The Correspondence of Admiral of the Fleet Lord Fisher of Kilverstone.* Vol III. London: Jonathan Cape, 1959.

Marjoram, P., J. Molitor, V. Plagnol and S. Tavare. "Markov Chain Monte Carlo Without Likelihoods." *Proceedings of the National Academy of Sciences* 100 (2003): 15324–15328.

Martin, George R. R. *A Song of Ice and Fire.* New York, NY: Bantam Books, 1991–2011.

Massie, Robert K. *Dreadnought: Britain, Germany, and the Coming of the Great War.* New York, NY: Random House, 1991.

Mastny, Vojtech. "How Able Was "Able Archer"?: Nuclear Trigger and Intelligence in Perspective." *Journal of Cold War Studies* 11, no. 1 (2009): 108–123.

McCann, Leo. "'Killing Is Our Business and Business Is Good': The Evolution of "War Managerialism from Body Counts to Counterinsurgency." *Organization* 24, no. 4 (2017): 491–515.

McCloskey, Donald N. "Counterfactuals." In *The World of Economics,* edited by J. Eatwell, 149–154. London: Palgrave Macmillan, 1991.

McCormick, John T. "The Hamlet Evaluation System – Reevaluated" (10 May 2021), https://storymaps.arcgis.com/stories/1c1bb536ec49 4f69815ab27feef47255, Accessed 4 January 2023.

McNamara, Robert S. *In Retrospect: The Tragedy and Lessons of Vietnam.* New York, NY: Times Books, 1995.

Megill, Allan. "The New Counterfactualists." *Historically Speaking* 5, no. 4 (2004): 17–18.

Miles, Simon. "The War Scare That Wasn't: Able Archer 1983 and the Myths of the Second Cold War." *Journal of Cold War Studies* 22, no. 3 (2020): 86–118.

Mueller, John E. "The Search for the 'Breaking Point' in Vietnam: The Statistics of a Deadly Quarrel." *International Studies Quarterly* 24, no. 4 (December, 1980): 497–519.

Mumford, David. "The Dawning of the Age of Stochasticity." In *Mathematics: Frontiers and Perspectives,* edited by V. Arnold et al. Providence, RI: American Mathematical Society, 2002.

National Archives and Records Administration. "HES71 and HAMLA Technical Documentation." Available at https://catalog.archives.gov/id/4658121, 2019.

Nolan, D. "Why Historians (and Everyone Else) Should Care About Counterfactuals." *Philosophical Studies* 163, no. 2 (2013): 317–335.

O'Mahony, Seamus. "Medicine and the McNamara Fallacy." *Journal of the Royal College of Physicians of Edinburgh* 47, no. 3 (2017): 281–287.

Office of the Deputy Chief of Staff for Military Operations, Department of the Army, A Programme for the Pacification and Long-Term Development of South Vietnam (PROVN) (March 1966), Vol. 1.

Office of the Historian. *Foreign Relations of the United States, 1964-1968, Vol. IV, Vietnam, 1966.* Washington, DC: US Government Printing Office, 1998.

Okun, Nathan. *Naval gun and armor data resource.* http://www.navweaps.com/index_nathan/index_nathan.php

Orange, Vincent. *Park: The Biography of Air Chief Marshall Sir Keith Park, GCB, KBE, MC, DFC, DCL.* London: Grub Street, 2000.

Osipov, M. "The Influence of the Numerical Strength of Engaged Forces on Their Casualties" (1915), trans. R. Helmbold and A. S. Rehm, *Naval Research Logistics* 42 (1995): 435–490.

Pearl, Judea and Dana Mackenzie. *The Book of Why: The New Science of Cause and Effect.* London: Penguin, 2019.

Pinker, Stephen. *Rationality: What It Is, Why It Seems Scarce, Why It Matters.* London: Penguin, 2022.

Poston, Tim and Ian Stewart. *Catastrophe Theory and Its Applications.* London: Pitman, 1978.

Poundstone, William. *Prisoner's Dilemma.* New York, NY: Random House, 1992.

Preble, Christopher A. "Who Ever Believed in the 'Missile Gap?'": John F. Kennedy and the Politics of National Security." *Presidential Studies Quarterly* 33, no. 4 (2003): 801–826.

Ramsey, Winston G. *The Battle of Britain: Then and Now.* 5th ed. London: Battle of Britain prints international, 1989.

Ranft, Bryan ed. *The Beatty Papers Vol 1: selections from the private and official correspondence of Admiral of the Fleet Earl Beatty.* Navy Records Society, volume 128. London: Scolar Press, 1989.

Reisch, George A. "Chaos, History, and Narrative." *History and Theory* 30, no. 1 (1991): 1–20.

Richardson, Lewis Fry. *Arms and Insecurity.* London: Stevens, 1960.

Richardson, Lewis Fry. *Statistics of Deadly Quarrels.* London: Stevens, 1960.

Rodger, N.A.M. *The Command of the Ocean*. London: W. W. Norton, 2005.

Roser, Max, Bastian Herre and Joe Hasell. "Nuclear Weapons". Published online at OurWorldInData.org. Retrieved from: https://ourworldindata.org/nuclear-weapons [Online Resource], 2013.

Roth, Paul A. and Thomas A. Ryckman. "Chaos, Clio, and Scientific Illusions of Understanding." *History and Theory* 34, no. 1 (1995): 30–44.

Ryall, Julian & The Korean Times, 'Moon Jae-in's Administration Faces Call to Investigate War Crimes Amid Rising Awareness of South Korea's Atrocities in Vietnam', *South China Morning Post* (18 April 2019), https://www.scmp.com/news/asia/east-asia/article/3006771/moon-jae-ins-administration-faces-calls-investigate-war-crimes, Accessed 4 January 2023.

Sarkar, Dilip. *Bader's Duxford Fighters: The Big Wing Controversy*. Worcester: Victory, 2006.

Schapiro, J. Salwyn. "Thomas Carlyle, Prophet of Fascism." *The Journal of Modern History* 17, no. 2 (1945): 97–115.

Schenk, Peter. *Invasion of England 1940: The Planning of Operation Sealion*. Berlin: Oberbaum, 1987; Engl. trans., London: Conway, 1990.

Scott, Len. "Intelligence and the Risk of Nuclear War: Able Archer-83 Revisited." *Intelligence and National Security* 26, no. 6 (2011): 759–777.

Sharp, Adm U. S. G. *Strategy for Defeat: Vietnam in Retrospect*. Novato, CA: Presidio, 1978.

Sharp, Adm U.S.G. & General W.C. Westmoreland, *Report on the War in Vietnam* (as of 30 June 1968) Washington, DC, US Government Printing Office, 1968.

Shultz, Richard. "Breaking the Will of the Enemy During the Vietnam War: The Operationalization of the Cost-Benefit Model of Counterinsurgency Warfare." *Journal of Peace Research* 15, no. 2 (1978): 109–129.

Silver, Nate. *The Signal and the Noise: The Art and Science of Prediction*. London: Allen Lane, 2012.

Sorley, Lewis. "To Change a War: General Harold K. Johnson and the PROVN Study", *Parameters* (Spring 1998): 93–109.

Sorley, Lewis. *Honourable Warrior: General Harold K. Johnson and the Ethics of Command*. Lawrence, KS: University Press of Kansas, 1998.

Sorley, Lewis. *Thunderbolt: General Creighton Abrams and the Army of His Times*. Bloomington, IN: Indiana University Press, 2008.

Sorley, Lewis. *Westmoreland: The General Who Lost Vietnam*. Boston, MA: Houghton Mifflin Harcourt, 2011.

Spencer, Herbert. *The Study of Sociology*. London: Henry S. King, 1873.

Spiegelhalter, David. *The Art of Statistics: Learning from Data*. London: Pelican, 2019.

Steel, Nigel and Peter Hart. *Jutland, 1916: Death in the Grey Wastes*. London: Cassell, 2003.

Summers, Harry G Jr. *On Strategy: A Critical Analysis of the Vietnam War*. Novato, CA: Presidio, 1982.

Sunstein, Cass R. "Historical Explanations Always Involve Counterfactual History." *Journal of the Philosophy of History* 10, no. 3 (2016): 433–440.

Taleb, Nassim Nicholas. *The Black Swan: The Impact of the Highly Improbable*. London: Penguin, 2010.

Taylor, Alan J. P. *Bismarck: The Man and Statesman*. London: Arrow, 1961.

Taylor, Alan J. P. *The Course of German History*. London: Hamilton, 1945.

Taylor, Alan J. P. *War by Time-Table: How the First World War Began*. London: Macdonald, 1969.

Tetlock, Philip and Aaron Belkin eds. *Counterfactual Thought Experiments in World Politics*. Princeton, NJ: Princeton University Press, 1996.

Tetlock, Philip and Dan Gardner. *Superforecasting: the Art and Science of Prediction*. London: Penguin, 2015.

Tetlock, Philip E. and Richard Ned Lebow. "Poking Counterfactual Holes in Covering Laws: Cognitive Styles and Historical Reasoning." *American Political Science Review* 95, no. 4 (2001): 829–843.

Tetlock, Philip, Richard Ned Lebow and Geoffrey Parker. *Unmaking the West: "What-If" Scenarios That Rewrite World History*. Ann Arbor, MI: University of Michigan Press, 2006.

Thayer, Thomas C. *War Without Fronts: The American Experience in Vietnam*. New York, NY: Routledge, 2019.

Townsend, Peter. *Duel of Eagles*. London: Weidenfeld and Nicholson, 1970.

Tuchman, Barbara. *August 1914*. London: Macmillan, 1980.

Tucker, Aviezer. "Historiographical Counterfactuals and Historical Contingency." *History and Theory* 38, no. 2 (1999): 264–276.

Turchin, Peter. *War and Peace and War*. New York, NY: Plume, 2007.

Turchin, Peter. "Arise 'cliodynamics'." *Nature* 454, no. 7200 (2008): 34.

Turchin, Peter and Sergey A. Nefedov. *Secular Cycles*. Princeton, NJ: University Press, 2009.

Warden, Col John A. *The Air Campaign: Planning for Combat.* Washington, DC: Brassey's, 1989.

Watt, Donald Cameron. *How War Came.* London: Heinemann, 1989.

Wells, Luke Benjamin. "The 'bomber gap': British Intelligence and an American Delusion." *Journal of Strategic Studies* 40, no. 7 (2017): 963–989.

Westmoreland, William C. *A Soldier Reports.* New York, NY: Dell, 1980.

Wirtz, James J. "Intelligence Please? The Order of Battle Controversy During the Vietnam War." *Political Science Quarterly* 106, no. 2 (1991): 239–263.

Wood, Andrew Jamie and Graeme J. Ackland. "Evolving the Selfish Herd: Emergence of Distinct Aggregating Strategies in an Individual-Based Model." *Proceedings of the Royal Society* B274, no. 1618 (2007): 1637–1642.

Wood, Derek and Derek D. Dempster. *The Narrow Margin: The Battle of Britain and the Rise of Air Power 1930–40.* London: Hutchinson, 1961.

Wyndham, John. *Web.* London: Michael Joseph, 1979.

Wynn, Kenneth G. *Men of the Battle of Britain: A Biographical Dictionary of the Few.* Barnsley: Frontline, 2015.

Young, Dave. "Computing War Narratives: The Hamlet Evaluation System in Vietnam." *Machine Research* 6, no. 1 (2017): 50–64.

Yuravlivker, Dror. "'Peace Without Conquest': Lyndon Johnson's Speech of April 7, 1965." *Presidential Studies Quarterly* 36, no. 3 (September 2006): 470–472.

Index

Note: Locators in *italics* represent figures, **bold** indicate tables in the text and page numbers followed by "n" denotes endnotes.